Essential Saltwater Flies

ED JAWOROWSKI

STACKPOLE
BOOKS

*For Michele,
the greatest catch of my life*

Copyright © 2007 by Ed Jaworowski

Published by
STACKPOLE BOOKS
5067 Ritter Road
Mechanicsburg, PA 17055
www.stackpolebooks.com

All rights reserved, including the right to reproduce this book or portions thereof in any form or by any means, electronic or mechanical, including photocopying, recording, or by any information storage and retrieval system, without permission in writing from the publisher. All inquiries should be addressed to Stackpole Books, 5067 Ritter Road, Mechanicsburg, Pennsylvania 17055.

Printed in China

First edition
Photos by the author
Illustrations by Dave Hall
Cover design by Caroline Stover

10 9 8 7 6 5 4 3 2 1

Library of Congress Cataloging-in-Publication Data

Jaworowski, Ed.
　Essential saltwater flies / Ed Jaworowski.
　　p. cm.
　ISBN-13: 978-0-8117-3459-2 (alk. paper)
　ISBN-10: 0-8117-3459-5 (alk. paper)
　1. Flies, Artificial. 2. Fly tying. 3. Saltwater fly fishing. I. Title.

SH451.J27 2007
688.7'9124—dc22
　　　　　　　　2007014288

CONTENTS

Preface iv

Chapter 1
Innovation and Variation 1

Chapter 2
Materials and Tools 3

Chapter 3
Techniques and Tips 13

Chapter 4
Baitfish Imitations 29
Bendback 30
Blonde 33
Bucktail Deceiver 35
Clouser 37
Deceiver 40
Half and Half 43
Hi-Tie 45
Jiggy 47
Seaducer 49
Siliclone 51
Slab Side 54
Surf Candy 56
Whistler 59
Wobbler 61

Chapter 5
Crabs and Shrimp, Eels and Worms 63
Chernobyl Crab 64
Del Brown's Permit Fly 66
Ultra Shrimp 69
Rabbit Eel 72
Worm Fly 74

Chapter 6
Topwater Flies 76
Bob's Banger 77
Crease Fly 79
Floating Minnow 82
Gurgler 84
Popper 86
Slider 89

Chapter 7
Bonefish Flies 91
Bonefish Special 92
Crazy Charlie 94
Gotcha 96
Horror 98
Mini Puff 99
Snapping Shrimp 101
Swimming Shrimp 103

Chapter 8
Tarpon Flies 106
Apte Tarpon Fly 107
Baby Tarpon Flies 109
Bunny Fly 110
Cockroach 112
Toad Fly 113
Whistler for Tarpon/Deep Water 115

Chapter 9
Fishing the Flies 118
Knots 118
Retrieves 121

Recommended Reading 123
Index of Fly Patterns 124

PREFACE

Saltwater fly fishing continues to attract new adherents. Some have fished extensively for saltwater game fish with spinning, casting, or trolling gear, and are interested in adding fly tackle to their arsenals. Generally, they have little or no fly-tying experience. Many others, coming from the ranks of freshwater fly anglers, and well versed in trout or warm-water fishing and fly tying, are anxious to make the transition to salt water. Then again, many saltwater fly fishers, familiar with certain species and the flies used to take them, want to test new waters, like the striped bass angler in quest of his first bonefish or the redfish and seatrout angler pursuing his first tarpon.

Essential Saltwater Flies is a primer for these three groups. If someone new to fly tying can master the basic designs covered here, he will have learned the fundamental techniques and how to handle the most important materials, and he will find it easy to move on to unfamiliar flies and the endless variations that he will encounter as he gets more immersed in the sport. Anyone crossing over from fresh water, and who is already familiar with basic tying procedures, will expand his horizons, learn to handle different materials, and tie flies often of larger proportions. Finally, the experienced saltwater fly fisher in pursuit of new species will find here basics to help him tackle his new challenges.

This isn't a master reference work or encyclopedia, cataloging the ever-growing list of individual patterns with their recipes. Many such works exist, and others will follow. In the 1950s, 60s, and early 70s, little was available in the way of saltwater fly tying instruction. Then came Kenneth Bay's wonderful little seminal work *Salt Water Flies* (Lippincott, 1972), with its step-by-step photo/text sequences. Bay showed procedures for eight basic flies that covered the majority of the fly fisher's needs. In concept, that's the approach I've taken, to give saltwater tiers a select list of all-purpose patterns based on a limited number of designs. Of course, saltwater fly fishing is no longer a fledgling sport. It has grown over the last two generations, in terms of species pursued, venues fished, new fishing techniques discovered, fly materials employed, and mostly in the cumulative knowledge amassed by saltwater fly fishermen. Naturally, therefore, I have had to expanded on Bay's offerings, although nearly all his basic designs are included here. I believe that armed with a small selection, chosen from among approximately three dozen flies described in these pages, any angler can feel confident and prepared to catch any target species under 90 percent of fishing circumstances and conditions.

The tier should find most of what he needs here: dressing recipes, images of finished flies, a bit of information about their evolution and development. Step-by-step photos and instructions constitute the heart of the book. Throughout the work he will find many tips for more creative tying, making adaptations, working with various materials, weighting flies, adding weed guards, as well as preferred knots for attaching the flies, retrieves, and other random techniques for fishing them. The tier need only supply the remaining, yet vital ingredient: practice. This handbook can get you started, take you to the next step, or provide new direction to your saltwater fly tying and fishing. Most of all, I hope it will help you get more enjoyment and satisfaction from your sport.

ACKNOWLEDGMENTS

Two individuals deserve special recognition. I owe a special debt of gratitude to my friend John Zajano, a versatile tier and talented angler who tied most of the flies for this book. His suggestions and ideas were indispensable, his patience and cooperation invaluable, his participation vital. Bob Popovics, a lifelong angling companion and one of the most creative tiers in the sport, tied the six "Pop Fleyes" designs included here. They are all simpler or improved versions of flies that he and I included in *Pop Fleyes* (Stackpole Books, 2001).

Ed Jaworowski
Chester Springs, Pennsylvania
2007

CHAPTER 1

Innovation and Variation

Individual saltwater fly patterns or dressings are seemingly endless; designs or styles, relatively few. By "pattern" I mean a specific dressing, typically limited by color, size, or material. Almost daily, season after season, thousands of new saltwater fly patterns are born in vises of tiers around the world. Yet, only occasionally does a new design come to light. "Design" is a conceptual thing. It represents a style or theme that invites variation of size, color, and material, and allows—indeed, can usher in—a whole field of creative adaptation and imitation. Bob Clouser's Deep Minnow, one of the world's most universal flies used in fresh and salt water, whether tied with long green and yellow bucktail, or short red and white Super Hair, still warrants the moniker "Clouser Minnow." The concept is the key. As opposed simply to endless variations, designs have pointed the fly tier's craft in new directions. In addition to Clouser's fly, Lefty Kreh's Deceiver, Larry Dahlberg's Diver, and Bob Popovics's Surf Candy immediately come to mind. They are innovative. From each of these examples, tiers have created hundreds, perhaps thousands of new flies. Most sport different names, and different creators lay claim to them, but changing the hook or color, or making some minor adjustment in technique amounts to variation, not innovation. It is important for the tier to appreciate this distinction. First, so as not to feel overwhelmed and bewildered by all the fly patterns he will encounter, recognizing that many are simply variations on a theme. Second, so that he can feel free to exercise his own creativity, and not feel restricted to a dressing that specifies exactly the color, material, or tying procedure. For years, that had been the bane of trout fly tying. Tiers searched in vain for a precisely dictated shade of dun hackle or olive dubbing to imitate a particular mayfly, as if all animals of a given species are precisely the same color. Saltwater flies are much less prescribed, as these photos illustrate.

These represent three versions of a tarpon fly called a Black Death. All are tied "Keys style," with all the materials at the back, but all use hackles and/or marabou feathers in different ways. The only similarity is a combination of black and red. Similar examples abound with other fly patterns.

Ironically, and despite this general fact, some bonefish and tarpon flies show little variation, yet do carry different names. Obviously, a Bonefish Clouser, a Crazy Charlie, and a Gotcha all represent similar design. Nevertheless, I have featured these individually because their identities, names, and dressings have become standard, and tradition dies hard.

Despite this anomaly, my general intention is to present a basic assortment of essential fly designs, as opposed to a wide mixture of random patterns. Nearly all the flies included in this work are standards or, in some cases, classics. Some newer, more contemporary flies have been included because they illustrate different tying techniques, or because they are destined to become standards. All are basic flies that every tier should know how to tie, because most new flies employ tying procedures and materials that these flies first used. Fortunately, and this should be encouraging, most are rather simple to tie, calling for familiar tying techniques and average dexterity, and frequently requiring only one or two different kinds of materials. Such is the case, for example, with the Seaducer, Blonde, Hi-Tie, and "Keys-style" tarpon flies. These flies will teach you to handle common materials like bucktail and saddle hackles. Some of these earlier designs too, although not so popular as they once were, have been included for just that reason: namely that familiarity with them will make anglers aware of the historic evolution of our sport and the rationale of its development. None represents a *tour de force,* tying simply for the sake of tying or showmanship.

Countless hundreds of other flies might have been included, many as good as or, for some purposes, perhaps better than some of those I've selected. But no one can have fished more than a small fraction of the existing patterns. Furthermore, no two anglers would select the same three-dozen basic flies. We all have our favorites, and certainly my own prejudices, shaped by my personal experiences, show through in each grouping. Still, for almost any species you target you can find here several proven and productive patterns. That is the purpose of this book. Credit to the true innovators often gets misplaced, and verification is often impossible. That is unfortunate, because their contributions have had a permanent and lasting effect on our sport. I have tried to accurately research the origins and evolutions of the flies I've selected and assigned them to their creators as best I could.

In each chapter and under each fly name, you will find the primary design illustrated in one specific pattern, with detailed, step-by-step photographs and tying instructions. You will also find variations of the basic designs. If you can tie one, by merely making minor modifications to the instructions, you can likely tie them all. Change the sizes or colors, even the materials, to suit your own desires or needs. Where it seemed especially appropriate, I have included notes and suggestions for fishing them, along with the species or conditions to which each design is generally best suited, lest we lose sight of the point of this craft.

I have never found a perfect system for classifying saltwater flies. Game fish regularly ignore our attempts to classify flies as species specific, i.e., bass, tuna, redfish, or bonefish flies. The same crab pattern can work for stripers or permit, and jacks eat bluefish flies. I once caught a juvenile bonefish in Belize on a large "tarpon fly" and a Bahamian permit on a "bonefish fly." Lefty's Deceiver, while initially designed as a striped bass fly, is about as universal a baitfish imitation as exists. You can make it short, long, slim, bulky, tiny, huge, solid, or multi-colored. It attracts countless species of small inshore or large offshore fish. Similarly, the Clouser Deep Minnow can be made in smaller sizes to fish for bonefish on sandy flats, weedless for redfish and snook, or weighted with large eyes (or even multiple pairs) to fish the deepest waters for striped bass. Still, since some organization is called for, I have presented the flies in categories that seemed logical, even if they are not completely consistent. Only two groups are aimed at specific fish: bonefish and tarpon. The largest group contains basic, generic baitfish imitations, adaptable to inshore or offshore use, for striped bass, bluefish, redfish, snook, weakfish, seatrout, snappers, jacks, yellowtail, bonito, and a host of others. Surface flies seemed to merit a separate chapter, as did specialized imitations of crabs, shrimp, eels, and worms.

Fly color is seldom dictated, except in a few cases like the Gotcha, Apte Tarpon Fly, and some others. Don't regard the colors of the examples shown as the most highly recommended or the most productive colors for that specific fly. I've tried to give the tier some "eye candy." We can only theorize why chartreuse succeeds on one outing, and blue and white the next. Most, like the Deceiver, Whistler, Seaducer, Bendback, Clouser, and Tarpon Bunny, readily encourage you to experiment with different color combinations. "Flash" is another matter of personal choice. If you feel adding some form of tinsel will enhance your flies, don't hesitate to include it. Better yet, tie some each way, with and without. Fish with them and decide if one is more productive for you than the other.

Fly tying is a matter of interpretation, making counterparts of natural forage as we imagine they will most appeal to fish. Just as no two artists will paint a horse exactly the same way, we shouldn't expect two tiers' versions of a shrimp or mullet to be exactly the same. Don't limit your thinking. You don't have to be a great tier to produce fish-catching flies. Use your creative instinct; apply your own ideas or requirements to your tying. Enjoy it. Don't change for the sake of change, simply to be cute, clever, or different. Experiment, within reason and always with a purpose, when tying flies and fishing them. To assist in this, note the "Additional Tying Notes and Variations" appended to each set of instructions. These will give you some idea of where tiers have taken various ideas, or where you might.

CHAPTER 2

Materials and Tools

IMPORTANT FLY TYING MATERIALS

The list of materials used by tiers is seemingly endless, and continuously growing. For that reason, the subject eludes complete or exhaustive treatment. I suggest you visit a few well-stocked fly shops, or study their catalogs and websites to become acquainted with the myriad of materials available. The items discussed and shown here will give some idea of the range and types of materials commonly used. Uses for most of these, and many others, will appear in the "Tying Tips" section of the following chapter and in the tying instructions for specific flies that follow.

Hooks

The hook is the most important ingredient of any fly. The hook should complement the fly design, but it must also penetrate readily and hold securely, without straightening or breaking. I am puzzled, and somewhat amused, by fly fishers who readily spend hundreds of dollars for rods or reels, and willingly fork over thousands of dollars on destination trips, then complain about spending fifty cents for a hook. Don't skimp on hooks. Understand the varieties and features of fly tying hooks, so that you can make better choices when it comes to hook selection. Tiemco, Owner, Varivas, Mustad, Partridge, Daiichi, Eagle Claw, and other companies make a wide range of excellent hooks suitable for fly tying.

Hook shanks may be "standard" length, longer or shorter. A 1/0 hook may be described at 2X long, meaning that its shank length is that of a 3/0 hook. Some flies, like the Jiggy and many poppers, require longer shanks. New Jersey master tier and fly designer Bob Popovics uses long shank hooks on a number of his fly designs. Famed California angler Dan Blanton favors an Eagle Claw jig hook for many of his flies (see "Whistler" flies). Most tuna anglers choose short shank hooks. They believe that the fish get more leverage with long shank hooks, allowing them to pull out or break. There are many bend shapes, too. The multitude of choices stems from tiers' preferences, and different shapes complement different fly designs. The O'Shaughnessy bend has been the most popular throughout the years, but on some fly designs it allows the materials to wrap around or foul readily. A Model Perfect bend may solve the problem. More recently, other bends, including different circle hooks, have become popular, and tiers have been busy modifying traditional fly designs to suit these unusual shapes.

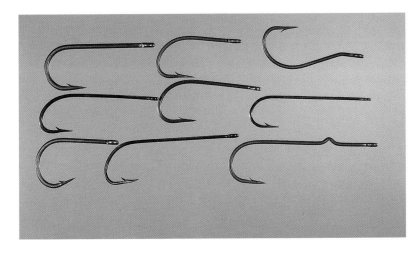

Every fly begins with hook selection, and tiers have a large variety of hook styles from which to choose.

Top row: lead, copper, and silver wire. Bottom row: clear mono, flat waxed thread, floss, and fine twisted thread.

Left to right: accelerator for cyano-acrylate, gel CA, brush-on CA, and three popular head cements.

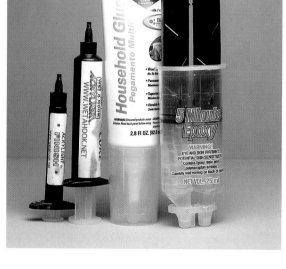

Clear acrylic, silicone, and 5-minute epoxy are some of the newest generation of fly tying materials.

The most common sizes used for saltwater flies run from 8, for small bonefish flies, to a hefty 4/0; some flies call for larger. The wire may be lighter or heavier, round in profile or forged flat to increase strength while limiting weight. Stainless steel is the choice of most tiers, but some prefer carbon steel or hooks plated with tin. One concern is that sharpening can remove the plating, exposing the metal to rust and corrosion. Long thin points can bend easily if they strike anything hard, and high barbs resist penetration. I fish barbless hooks exclusively. They do less damage to fish and are safer for anglers. I simply crush the barb flat with pliers. If you prefer to use barbed hooks, choose hooks with small "micro barbs" that penetrate readily but hold well.

Threads

As with most tying materials, thread choice largely depends on personal preference. I use various threads, but I am partial to Uni-Thread in 3/0 or finer 6/0, especially for smaller flies. My tying friend John Zajano prefers Danville, flat-waxed nylon floss; hence, most of the flies in this volume are tied with that. It's strong enough for most saltwater tying operations, comes in a wide range of colors, and doesn't create appreciable bulk. Bob Popovics ties almost exclusively with clear monofilament tying thread. Some tiers use white thread for nearly everything and simply color the last few inches with a permanent marker, reasoning that most often the color shows only on the head of the finished fly. Try a few different thread types and use what works best for you.

Glues and Cements

Hard as Hull, Dave's Flexament, Pharmacist's Formula, Loon Water-Base Head Cement, and Hard As Nails nail polish (available at cosmetics counters), are some of the most commonly used head cements. Look for products that are clear, penetrating, and dry hard. A brush attached to the cap makes application easier.

Cyano-acrylate (commonly referred to as CA), with such band names as Krazy Glue and Zap-A-Gap, is a quick-bonding cement with many applications. Loc-tite CA in gel form comes in a small, convenient squeeze bottle, and Krazy Glue is available with a brush-on applicator. A tiny spritz of an accelerator or "kicker," available in pump bottles, will harden this glue instantly. Note: exercise care when using CA. Acetone or paint thinner may be needed to release fingers that inadvertently get bonded. Read instructions carefully on these and all chemical products, and only use them in well-ventilated areas.

Bob Popovics's "Pop Fleyes" designs have made epoxy, such as Devcon clear 5-minute epoxy, a tying staple for the past twenty years. Use this two-part cement, composed of

Deer tails, commonly known as "bucktails," are one of the most useful materials for tying saltwater flies.

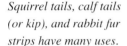

Squirrel tails, calf tails (or kip), and rabbit fur strips have many uses.

epoxy and a hardener, to shape and model fly bodies, to protect thread wraps, and to add durability to flies. Flex Coat, an epoxy commonly used by rod builders to protect wraps, is a bit thinner and retains some flexibility. Thirty-minute epoxy takes longer to gel and set, but it has a lesser tendency to yellow. One-minute epoxy is fine for coating heads, but allows too little working time for sculpting most fly bodies. Placing epoxy components under a warm lamp for a short time will make them flow more readily and easier to work with.

Acrylic plastic, similar to that used in dental reconstructions, is gaining in popularity as an alternative to epoxy. Bob Popovics, "Mr. Epoxy" himself, now uses it in lieu of, or as an alternative to, epoxy for many of his fly designs. It doesn't have the weight of epoxy, which may be an advantage or disadvantage, depending on your needs, but it remains clear and doesn't yellow as epoxy does. It is hard, doesn't run or chip, and is easy to apply. Tuffleye brand, manufactured by Wet-A-Hook Technologies, comes in handy syringes. Also unlike epoxy, acrylic doesn't harden until you expose it to a "bluelight," relieving the angst that sometimes accompanies epoxy use. Exercise extreme caution when using any such lights; never shine them toward your eyes.

Silicone, a rubber compound, such as GE or DAP caulking compound, has found its way to tying benches. Softex is another rubber product used to coat fly bodies. Simply dip the fly into the jar. Instructions for using silicone, epoxy, and acrylic are found throughout the text in the appropriate places.

Natural hairs and furs

The hair from deer tails is one of the most widely used natural materials for saltwater tying. The term "bucktail," as opposed to "deertail," has been sanctioned by use for so long that most tiers use that designation, whether the tail comes from a buck or doe. Available dyed in any color imaginable, bucktail has natural taper, buoyancy, and excellent action in the water. Hair toward the base of the tail is progressively more hollow and flares when compressed with thread. Hair toward the tip of the tail is more manageable and suited to streamer flies. Deer and elk body hair can be spun and flared to make full fly bodies.

Certain other tails can be useful for fly tiers as well. Calf tail has some of the same uses as bucktail but is short, crinkly, and less buoyant. I find it excellent for baby tarpon flies and small versions of Lefty's Deceiver. Gray squirrel and red fox squirrel tails have soft fibers and make nice bonefish flies and small Clousers, or collars around Deceiver flies.

Strips of rabbit fur make excellent streamers. The fur is lifelike and breathes enticingly with the slightest movement. The Zonker, a trout fly, first popularized the use of rabbit, and I employ it on the Rabbit Eel, Tarpon Bunny, Rabbit Surf Candy, and Fur Strip Clouser. These Zonker strips have the

Saddle pads contain long, tapered feathers in a wide range of sizes.

Strung saddle feathers are all approximately the same size.

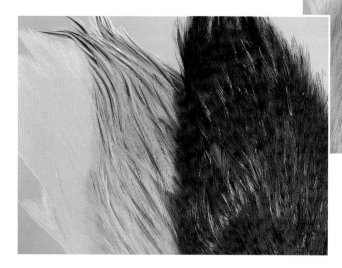

Necks, or capes, have smaller feathers. While more useful for trout flies, they have uses in salt water tying as well.

fur running parallel to the strip of hide. On crosscut strips, the fur runs perpendicular to the leather. These are wrapped in spiral around the hook shank, rather than trailing lengthwise. Icelandic sheep and yak hair provide fibers up to 12 inches in length, for really huge saltwater flies, and sheep fleece makes a plush body, as called for by Popovics's Siliclone flies.

Feathers

Saddle hackles are indispensable to saltwater tiers. These long feathers come from the back, or saddle, of chickens. Some feathers from the saddle are long and sleek; others, wide and webby. Short, wide spade hackles come from the base of the saddle. Each has uses. You can buy saddle hackles on the skin, or sorted and strung according to size. An entire saddle patch provides a wide range of sizes, but fewer of any one size, while strung hackles are all the same approximate size. Neck, or cape, hackles come from the area around the chicken's head and neck. They are shorter than saddle hackles, and while highly prized for trout flies, they are less used in saltwater tying, but some flies, such as Seaducers and Keys-style tarpon flies, do call for wide, curved neck hackles, and some flies call for hackle points from smaller feathers. Schlappen, wide, webby feathers more commonly seen on Las Vegas showgirls

MATERIALS AND TOOLS 7

Left to right: marabou, ostrich, and peacock herl add life and movement to many flies.

Chenilles and yarns are important fly body materials.

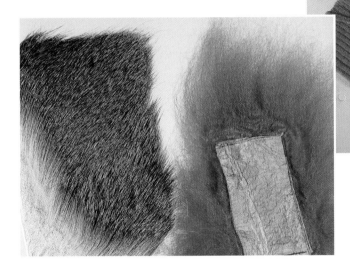

After being tied in, deer body hair and sheep fleece can be trimmed to provide a wide variety of body shapes.

than on tying benches, can be 10 inches or more in length. Certain patterns call for these or other less common feathers.

Several other feathers have useful applications. Marabou feathers, originally from the African marabou stork, today come mostly from young turkeys. No other material has the life of marabou, fluffing and breathing in the water. Flues from ostrich plumes and peacock herl are predominately used as toppings on flies to complement other materials.

Body materials

Abundant chenilles and yarns, natural and synthetic, originally produced for carpet and other industries, have found their way into fly-tying supply catalogs. Try various ones as substitutes and alternatives for the materials called for in traditional fly dressings. You may come up with some novel ideas.

Some important synthetics

The introduction of synthetic hairs and fibers represents the most important contribution to the growth of modern saltwater fly tying. They make the tying of larger flies easy, often with little excess bulk or weight. The products available are legion. Ultra Hair, Super Hair, Unique Hair, Fishair, Big Fish Fiber (Hairabou), Polafibre, Craft Fur, and Kinky Fiber are just a few of the brands on the market. DNA and some other makers

8 ■ ESSENTIAL SALTWATER FLIES

A small sampling of the many modern fly tying synthetics: (left to right) DNA Frosty Fish Fiber, Fluoro Fiber, Fishair, Ultra Hair, Super Hair, and Kinky Fiber.

(left to right) Krystal Flash, Angel Hair, Comes Alive, Flashabou, and Sparkle Flash each give different flash effects when combined with other natural or synthetic materials.

Subtle flash effects can be produced by (from left) hollow braided piping, Cactus Chenille, Estaz, and tinsel chenille.

have fibers with unique light-catching properties, in addition to fiber and flash blends, such as Holo-Fusion fiber and Farrar's Flash Blend. Unlike natural hairs, most of these have no taper, so we achieve tapered fly silhouettes by trimming the hairs to various lengths. They often replace natural hairs on Deceivers, Clousers, Jiggies, Surf Candies, and other common flies. Search out fibers that shed water readily, instead of absorbing it. You can custom blend synthetics to your own specifications by mixing fibers of different colors.

Tinsel and other flash
Years ago, only metallic tinsels were available. These rusted or otherwise corroded and just couldn't withstand the rigors of the saltwater environment. It is no accident that many early saltwater flies lacked flash, which we now consider almost essential. Flash may come on spools or in hanks; short or long; flat, round, or oval; holographic or Mylar; and in endless colors. Flashabou is my all-time favorite tinsel. Tightly twisted strands of Flashabou produce Krystal Flash, which has more body, with a different and subtler flash effect. Let several strands of either extend slightly longer than your hair wing for a "flashtail" effect. Mirage, Angel Hair, Comes Alive, Fluorofibre, Sparkle Flash, and others all yield different effects and give tiers endless opportunities to experiment. Pearlescent braids, and chenilles like Estaz and Cactus Chenille, offer even more options.

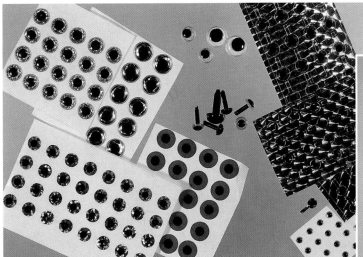

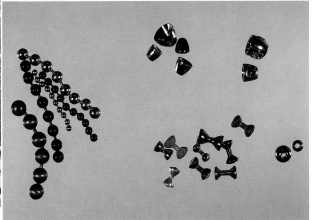

A small sampling of the many styles of eyes used by fly tiers.

Weighted eyes and heads for saltwater flies come in many sizes and colors: (clockwise from left) bead chain, cone heads, Jiggy heads, metallic beads, and dumbbell eyes.

Clear Stretch Tubing, Larve Lace, and round rubber legs.

Permanent marking pens have a variety of uses for saltwater fly tiers.

Eyes

Eyes on flies can serve several purposes. Bead chain and lead or tungsten dumbbell eyes are used principally for weight and to provide jigging action, as on Clousers, Whistlers, or Crazy Charlies. Three-dimensional glass or plastic doll eyes add some weight but are mainly intended to add realism. The most useful decorative eyes are prismatic, self-sticking eyes. These adhesive-backed eyes can be peeled off a sheet and pasted to a fly or popper. Although they have some glue on the backing, a spot of cement assures they won't come off readily. Some also have a small tab that can be secured with tying thread. Eyes can also be painted onto feathers.

Weight

In addition to lead or other dumbbells, brass or tungsten cone heads can be added to many patterns and are an essential element for Jiggies. A special, modified cone head, called a "Jiggy head," has been specifically designed for this fly by Bestco Products. Lead, or lead free, wire can also be used underneath the body of many flies to help sink them.

Miscellaneous

Rubber bands and silicone "Sililegs" and "Crazy Legs" can add life to shrimp, crab, and squid flies. V-Rib, Larva Lace, and other stretchy tubing make nice translucent bodies when wrapped over tinsel bodies. Clear monofilament line can also be used. Sheet foam of various sorts, such as that used on the Crease Fly, is useful to tiers, as well as pre-shaped popper heads and bug bodies. Permanent marking pens marketed by Sharpie, Prizm, Prismacolor, and Pantone, available from fly shops or stationery stores, are useful for coloring feathers, hairs, and other materials and for marking gills.

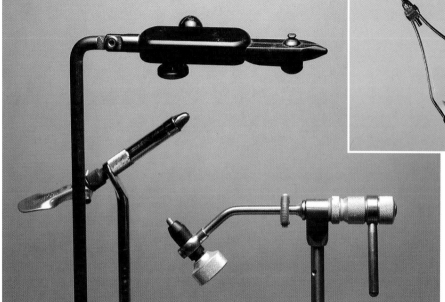

Bobbins (shown here with metal and ceramic tubes), a bobbin threader, and a bodkin are among the most basic tools.

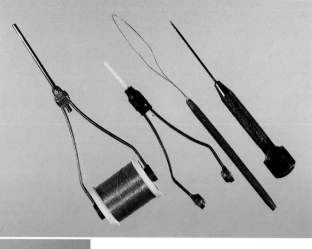

Many vises, from quite inexpensive to the priciest, will serve you well. Choose based on hook-holding ability, ease of operation, durability, and variety of features.

BASIC TOOLS

Fly tiers use a wide variety of tools, most especially designed for specific operations, and tackle catalogs and fly shops are replete with them. Many are a matter of personal preference. Here are some of the most commonly used. The "Tying Tips" section of the next chapter explains how to use some of these, and the tying instructions in later chapters demonstrate others in action.

Vise

The primary function of a vise is to hold the hook securely. Surprisingly, not all do this well. Before selecting a vise, test its hook-holding ability with a series of hook lengths and sizes, as well as its ease of operation and adjustment. A cam-operated lever that quickly engages and releases the jaws will prove a blessing in the long run, and don't overlook the range of hook sizes that the jaws can accommodate. Some vises don't handle larger hooks so well. Most saltwater flies run from about size 8 to 4/0. If you plan to use the same vise for freshwater tying, you'll need one capable of handling much smaller hooks. Consider thickness of the hook wire, too, when judging your needs. A round wire hook will be appreciably thicker than a forged hook, although the two are nominally the same size, and a vise that opens wide enough for one may not hold the other.

Modern vise makers offer a host of accessories for their products: interchangeable jaws designed for specific flies, high speed cranks, bobbin holders, material clips, extension arms of various types, and so on. You can spend $40 or the better part of a $1000 for a top-of-the-line model. Obviously, the more bells and whistles, the more money. Few of these add-ons are necessary for general fly tying. However, I strongly recommend you give attention to two features: First, a full, 360-degree rotating vise will make the time you spend tying enjoyable. Consider products by established manufacturers like Renzetti, Dyna-King, Regal, and a few others. Jaws that you can turn without losing their grip allow you to examine the fly from different angles and perform many operations more easily. Second, decide whether you want a C-clamp that fastens to the edge of a table or a heavy pedestal base that you can position anywhere. The latter is now far more popular. You can set the vise back from the edge of the table, giving yourself more working room and allowing you to rest your arms against the table.

Bobbin

Strictly speaking, a bobbin is the spool on which thread or yarn is wrapped. In fly-tying parlance it has come to mean the holder for the thread spool. Of the many designs that have been used, the simple wishbone bobbin is far and away the most popular. It uses spring tension to keep pressure on the thread. I recommend those with ceramic tubes. Some with metal tubes develop grooves or burrs that can cut your thread. You can run the thread through the tube with a bobbin threader or simply insert the end of the thread into the bobbin tube and suck it through. Most tiers find it convenient to have several bobbins prepared for different threads, to save time changing threads.

MATERIALS AND TOOLS

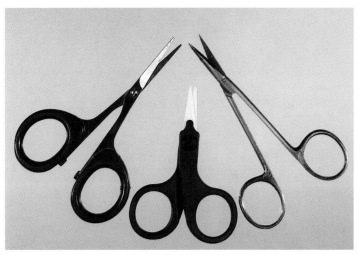

A pair of straight, general-purpose scissors with a serrated blade is essential. Scissors with ceramic blades, and fine point iris scissors are some of your many options.

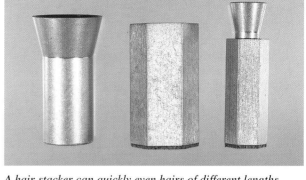

A hair stacker can quickly even hairs of different lengths.

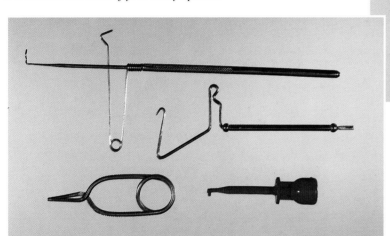

A hair packer or pusher performs functions that could be done by hand, but many tiers prefer to use this simple tool when packing deer hair tightly. Two types are shown here.

The whip-finishers at the top and the hackle pliers represent tools that facilitate certain tying chores.

Bodkin

A bodkin, often called a "dubbing needle," since picking out fur dubbing on trout flies was its principal early function, is an indispensable tool. With it, you can apply head cement, sculpt epoxy bodies, straighten hair, untangle fleece, and perform other sundry tasks.

Scissors

Don't skimp on scissors. You'll use them endlessly and if they don't function well, you won't enjoy tying. Get scissors specifically designed for fly tying, and consider at least two pairs. The first should have at least one serrated edge blade. Slick synthetic hair like nylon will slide between the blades when you attempt to cut it. A serrated blade holds the material lightly in place to facilitate cutting. Whether you purchase straight or curved blades is largely a matter of preference, but you'll find that you can perform some operations, like trimming deer hair, better with curved blades. For delicate work, get a pair of curved surgical scissors with very fine points. Make certain that the finger loops on any scissors you purchase are large enough to easily slip them on and off.

Additional and optional tools

You can tie off the finished fly by hand, as described in the next chapter, or use a tool designed for this purpose. I think every tier should know how to do both. Occasionally you will need to whip-finish thread while getting around bulky or protruding material, and most whip-finish tools simply don't perform well in such situations. You'll have to perform the operation by hand. Griffin, Matarelli, Renzetti, Thompson, and others, make whip-finishers, but they all work slightly differently, so follow the instructions that come with the tools. The Web also has plenty of videos and instructions on how to use them. Search under "whip-finishers." Hackle pliers are indispensable for tying trout flies, but not nearly so important for saltwater fly tying. Nevertheless, many tiers find they give added control when winding hackles, especially smaller ones. Larger hackles can readily be wrapped by hand. Similarly, hair stackers and packers, drying wheels, and other accessories make specific tying operations easier or contribute comfort and convenience to tying activities.

An electric or battery-powered wheel that rotates slowly allows you to turn a dozen or more epoxy flies simultaneously, as opposed to turning them by hand one at a time. After you apply the epoxy, stick the fly into the wheel until the epoxy gels and sets. You can make one from a slow rotisserie motor, or buy one ready to use. Note that a drying wheel is only used for epoxy flies that are round in cross section.

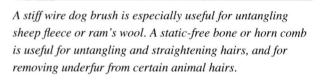

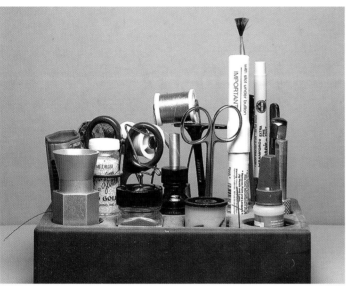

A stiff wire dog brush is especially useful for untangling sheep fleece or ram's wool. A static-free bone or horn comb is useful for untangling and straightening hairs, and for removing underfur from certain animal hairs.

Buy or make one or more tool caddies to organize tools and keep them readily available.

Drying racks for finished flies as well as hangers for organizing tinsels are convenient accessories. You can purchase them in fly shops or easily make them.

Comfort considerations

Spend some time preparing a tying site where it is convenient and comfortable to work. A good light placed over your vise, one specifically designed for close manual work, will ease eye strain and make you more comfortable. If your eyes aren't what they used to be, consider a swing-away magnifying lens, or at least leave a pair of "cheater" reading glasses on your tying table. Finally, a comfortable chair at your tying area, as at your computer or desk, is too often overlooked. You won't enjoy tying if your shoulders are hunched or if you are not sitting in a natural position, and you probably won't realize the cause.

CHAPTER 3

Techniques and Tips

BASIC TYING TECHNIQUES

Attaching thread to the hook

When I attempted to tie my first fly, a half century ago, I was stymied by the first move: how to attach the thread to the hook shank. A clinch knot, one of the few I knew, failed to secure it, as did a series of original experiments involving glue. I recall how stupid I felt when someone showed me what I suppose I should have been able to deduce for myself; it seemed so obvious after the fact. It is utterly simple to do, but I include the procedure here (using heavy cord in lieu of tying thread on a bobbin) to spare any beginning tier my embarrassment. Every fly begins with this procedure and some even call for tying off and later reattaching the thread. Just hold the end, lay the thread across the hook shank, and wrap it around the shank, making several wraps over itself; then trim the tag end with your scissors once the thread is securely attached.

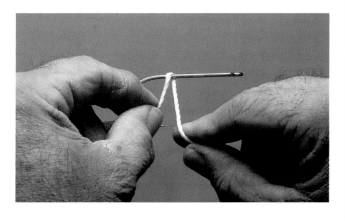

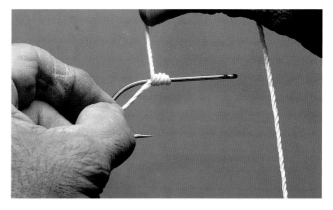

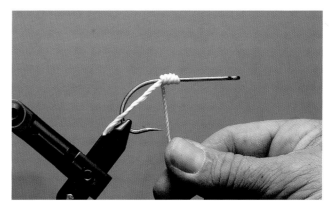

Whip-finishing the thread

Just as every fly starts with the simple step of attaching the thread, every fly finishes with tying it off. Some tiers use a series of half hitches, but the whip-finish is more secure and neater. Whip-finish tools are available (see "Tools" section in the previous chapter), but every tier should know how to whip-finish by hand. Sometimes, a whip must be made over bulky material and the tools are less convenient and even awkward to use. I cannot overemphasize the importance of mastering this procedure. It will seem complicated at first, perhaps impossible, but nothing is more important for you to learn. Practice, practice until you can do it instinctively. Basically it involves moving the index and middle fingers back and forth as you move your hand in a circular motion around the hook. Note: there are several ways to whip-finish by hand; this is one of the best.

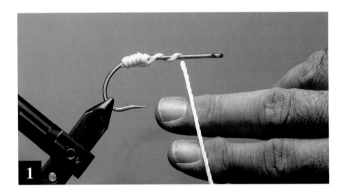

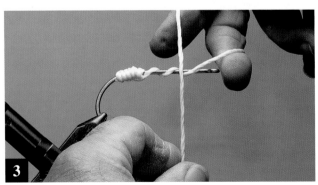

TECHNIQUES AND TIPS ■ **15**

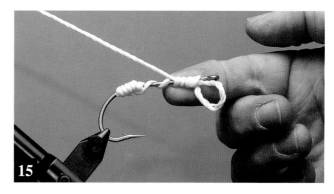

As stated earlier, there are several ways to whip-finish a fly by hand. Unfortunately, all are rather difficult to explain or illustrate with still photos. If you search for "whip-finish" on your computer's search engine, you will find several sites that demonstrate the various techniques. A number of tying videos and DVDs also illustrate them. Practice along with the film. It is vital that you master some form of the whip-finish.

SOME USEFUL TYING TIPS

The various procedures and tips that tiers collectively use could easily fill a book. I've included here a small selection that addresses common problems. They will help you in fashioning the flies in this book. Others appear throughout the text accompanying specific flies. For clarity, I illustrate them here in isolation, as modular components, rather than within the tying sequence where they would normally occur. I am especially indebted to John Zajano, who tied most of the flies in this book, for introducing me to several of these. Practice them on a bare hook until you can perform them comfortably and efficiently before employing them on the actual flies. It will make your fly tying that much more enjoyable.

Applying self-sticking, prismatic eyes

Fold the eyes in half, so that you put a good crease in the center of the row of eyes. This will prevent them from pulling away and springing straight when you attach them to the head of a fly or to a popper body.

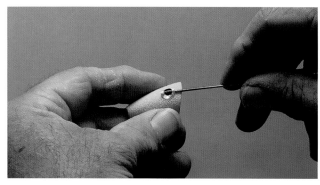

Pick the eye off the paper, transfer it to the fly, and press it firmly in place. Unless you are going to coat the eye with head cement or epoxy, put a small dab of CA glue or Goop on the fly before setting the eye in place.

Tying in tinsel

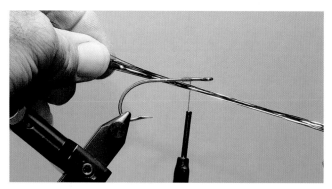

Start with half the amount of flash you want, since the tinsel will be doubled. Holding the strands of tinsel between your hands, bring them from below the hook.

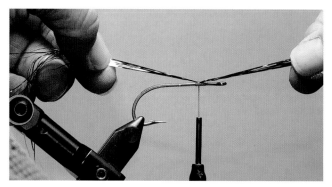

Slide the tinsel up on top of the hook under the thread.

Tie the tinsel down with several wraps of thread.

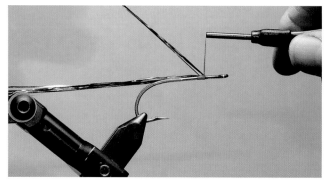

Fold the other half of the tinsel back and secure it with several more turns of thread.

Adding a tinsel throat (beard)

This touch of red adds additional flash. Perhaps it replicates the gills of a baitfish, but it undoubtedly enhances the appeal of many flies to both tiers and fish. Dealing with short pieces of tinsel can be difficult, and by cutting off short sections from strung tinsel, you waste material and end up with various lengths, some of which are useless. This procedure gives you control and avoids waste.

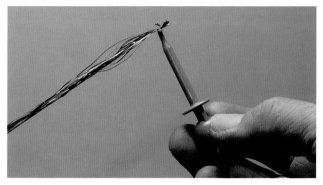

Cut about a dozen strands of Flashabou, Krystal Flash, or other tinsel, and grasp them in a wire clip. Wet them slightly for easier handling and control.

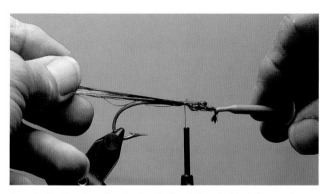

Attach the tinsel under the hook shank just behind the eye.

Pull the clip forward until you have the desired length of throat extending back. Trim the excess and fasten the tinsel securely with thread. The rest of the tinsel remains in the clip ready for the next use.

Securing material to the hook shank

Tying in almost any material—synthetic or natural hair, fur, chenille, hackles, etc.—and keeping it firmly in place, involves controlling thread pressure. This technique is commonly known as "the pinch."

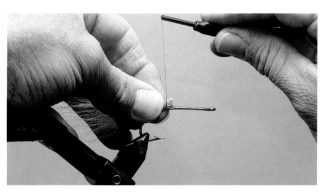

Hold the material directly over the hook shank and bring the thread up on the near side between your thumb and the hook.

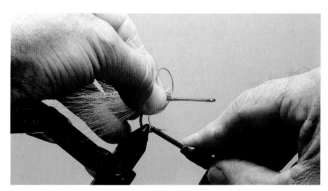

While keeping the material pinched firmly between your thumb and index finger, form a loop in the thread and work it between your finger and the material on the far side of the hook.

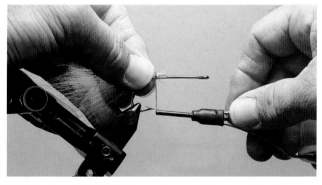

Pull the thread straight down and repeat the operation. Now pull down tightly on the thread while holding the material on top of the shank.

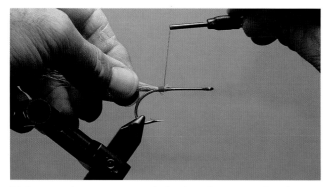

Make several more wraps with the thread to assure that the material is firmly in place. If you want hair evenly distributed around the shank, make two semi-taut wraps, then loosen your finger and thumb hold, and gently slide the material around the shank as you pull the thread tight with the bobbin.

Tying in chenille

Before attaching a piece of chenille, scrape the fuzz from the end of the chenille with your thumbnail and index finger to expose the core, and attach this core to the hook with your thread. This will eliminate much bulk and make neater flies.

Mixing epoxy

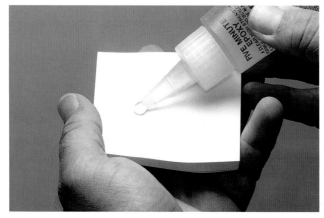

Squeeze a small amount of epoxy onto a note pad or piece of cardboard. Often no more than a dime size spot is required.

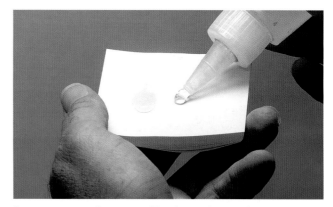

Squeeze out equal amount of hardener close to the epoxy.

Mix by folding the two parts together with a bodkin. Don't stir too vigorously or you will create bubbles in the mixture. Continue to turn the fly by hand, or with a drying wheel, until the epoxy gels and sets. Test the epoxy on the paper with your bodkin to determine when it is no longer fluid and you can cease rotation.

Painting lead or tungsten eyes

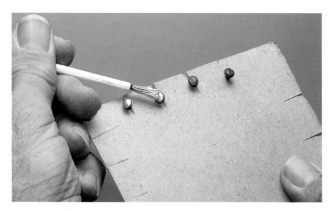

Cut small slots in a square of thin cardboard, and insert the eyes in the slots. Paint the iris of each with model or hobby paint. I prefer silver for most saltwater flies, but yellow, red, and white are also popular. (Don't forget to turn the card over and paint the eyes on the other side!)

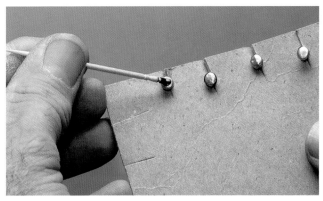

Using a toothpick, add a black pupil to each eye. Optionally coat the paint with a thin coat of epoxy for durability and to avoid chipping.

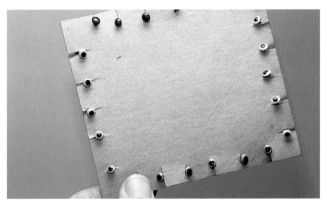

Prepare a bunch in advance, and store them in a plastic box, sorted by size and color, so that they will be ready when you are to tie a bunch of flies.

To even bucktail or other natural hair

To take advantage of the natural tapers of hair and to give the fly symmetry and shape, the hair ends should be reasonably even, but not so squared off as to resemble a paintbrush. You can do this by hand.

Cut a bunch of hair from hide, skin, or tail with your scissors.

Notice that many fibers are of different lengths.

Hold the butts in one hand, and pull out the longest tips with the other.

Now replace them in the bunch, i.e., combine all the hairs again, with the tips close together. You may have to repeat this several times.

The ends are together and the hair is now ready to be tied to the hook.

Using a hair stacker/evener

This is an alternate method many tiers use to even hairs. It's generally faster than working by hand. Again, however, avoid making a paintbrush, which can happen all too easily when using this tool. You want a little variation in hair lengths.

Cut a bunch of hairs from the hide and place them, tips down, into the stacker. Note that the butts are even. Tap the stacker on a hard surface several times. This will make the hairs slide so that the tips are even.

Remove the inner tube of the stacker. Note that the butts are now uneven, but . . .

. . . the tips are even.

Tying in thick bunches of hair

Often when working with large or thick bunches of hair, natural or synthetic, it is difficult to maintain enough tension on the bundle so that the hairs don't pull out and allow the fly to come apart. Here's how you can beat the problem.

Take your thread through only half (or a third) of the hair on the first wrap. It will also help to put a drop of head cement on the butt ends of the hair before starting. When the thread is pulled tight, this will force the cement between the hairs and make a more secure foundation.

Make your second (or third) wrap over the entire bundle of hair.

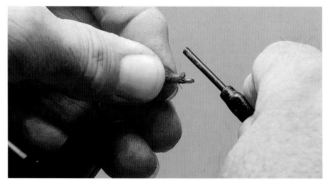

Make additional tight wraps until the hairs are firmly bound in place.

Spinning and packing deer hair

Cut a bunch of deer or elk belly or body hair from the hide. Optionally, you can cut across the tips so that the butt and tip ends are squared off.

Draw the hair through a static-free comb to remove short hair and underfur.

While holding the hair tightly in a pinch, make two or three wraps of thread, but don't pull the thread tight yet.

While steadily pulling down tightly on the thread, gradually release the hair from your pinch. The thread pressure will crimp the hollow hair and cause it to spin around the shank so that it is evenly distributed.

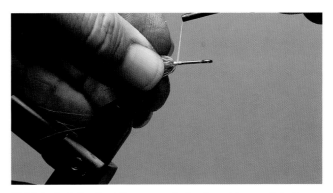

Hold the hair back while you work the thread forward, in front of the hair, and make a few more wraps.

With a hair pusher, push the hair tightly back, while holding the hair from the back with the thumb, index, and middle finger of your left hand. Keep repeating the five above steps until you have as much hair on the shank as needed. Practice packing the hair tightly after each bunch is tied in.

Curving or straightening hackles or herl

Tarpon flies, Seaducers, Deceivers, and others sometimes call for curved hackle feathers. Taking them from the same place on a neck assures that they will be the same length and width, but often they curve in different directions when tied onto the hook. Also, before wrapping a hackle collar, it can help to curve the hackle stem in the direction you intend to wrap, to make the wrapping easier. It's a simple matter to adjust the curve of any hackle stem. You can curve straight hackles or straighten curved hackles in a snap with this technique.

Note that this hackle is basically quite straight.

While holding the butt in one hand, draw the hackle stem between your bodkin and your thumb as shown.

Like magic, the stem is now curved. Obviously, you can remove the curve and straighten the hackle in the same way, by placing the bodkin on the opposite side of the stem when you draw the hackle through. You can use the same technique to add a nice curve to peacock herl used as a topping over a Deceiver or other fly, so that it follows the contour of a hackle or hair wing. By doing this after the herl is tied in, you can easily judge how it conforms to the rest of the fly's shape.

Tie in a small bundle of herl. Notice that it is straight.

Hold the tips in your left hand while gently stroking back and forth on the stems with a bodkin.

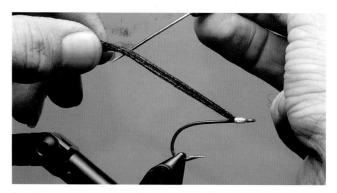

Starting at the bases, work your way to the tips.

When you release the herl, notice that the flues have now assumed curves.

Adding weight to a fly

Method 1

You can get a fly deeper in the water column by fishing it on a sinking line. Sometimes, however, adding weight directly to the fly, either as an alternative to the heavy line or in combination with it, can be effective. Here's how to weight your flies with minimal disruption to their shape and form and to your tying procedures.

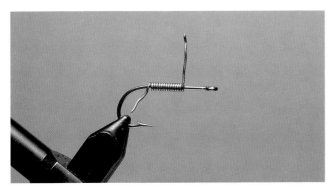

Wrap several turns of lead or tungsten or copper wire around the shank. Pinch off the ends with your thumbnail and index finger.

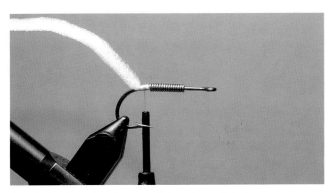

Attach the thread at the rear of the wire and tie in a narrow strip of foam packing material. Bring the thread to the front of the lead wraps.

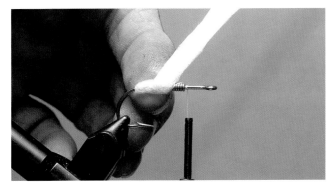

Wrap the foam tightly over the lead wire.

Tie down the foam ahead of the wire and trim the excess.

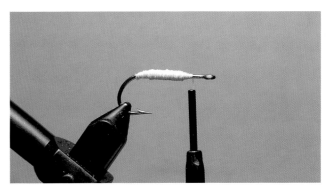

Wrap the thread tightly over the foam, compressing it and popping the air pockets. This will leave you with a smoother surface to work on than the bare wire when completing the fly.

Method 2

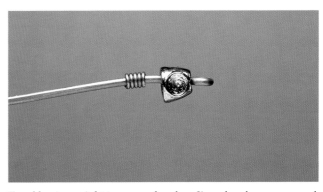

To add extra weight to a cone head or Jiggy head, wrap several turns of lead or other wire around the hook shank; then push the coil of lead inside the head and place a drop of CA cement inside to hold it in place.

Method 3

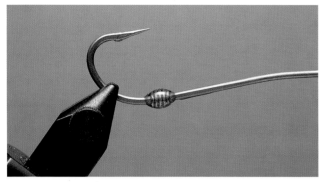

To weight a Bendback shank and give better keeling to the fly, wrap several turns of lead wire around the shank at the bend, and coat the wraps with epoxy.

Wetting feathers and tinsel for better control

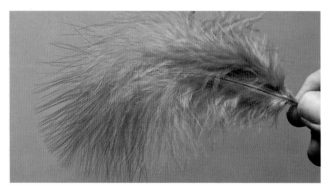

A dry, light, fluffy marabou feather like this can be difficult to manage, and it's difficult to avoid tying down the wavy flues inadvertently.

Draw the feather through a wet paper towel, cloth, or sponge several times.

The moist feather is easier to handle and control when tying it in.

Dry and wet hackles. You can do the same with hackles. Many times it makes them easier to work with during tying operations. Many of the flies tied in this book show wet hackles that will dry and fluff again by the time the flies are tied. Do the same with tinsel to keep it out of your way while dealing with other materials.

Here's another option. Place a small wet sponge on a dish or in a shallow bowl, with about an eighth of an inch of water in it. Moisten your fingers and draw them over hackles or press the feathers down and draw them over the sponge. When the sponge gets dry, flip it over and use the wet side. Do the same with strands of tinsel to keep them out of your way while dealing with other materials.

Note: *Some tiers wet the finished fly in an attempt to see what it will look like in the water, but this is not a good idea. If you wet the hair or feathers and stroke them on the fly, the fibers or flues will cling together, but once the fly is immersed, the hairs will separate again, and the fly will look just as it does when dry.*

Clearing the eye

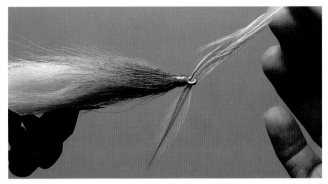

It's easy to get head cement or lacquer into the eye of a fly when cementing the head. To wipe it away before it dries, pull a saddle hackle feather through the eye. The feather will collect the excess cement.

Attaching a "stinger hook"

Occasionally fish will repeatedly strike short, grabbing the tail feathers of a fly, resulting in missed strikes. This can occur even with species that normally grab bait by the head in order to swallow it whole, like striped bass. Other species like bluefish and barracuda generally attack from the rear, but sometimes they grab merely the end of the wing. One solution is to use a trailer or "stinger hook" incorporated into the construction of the fly.

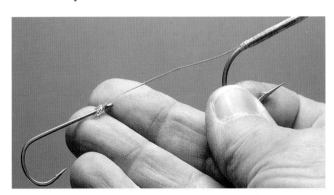

Before starting the fly, run a piece of stout monofilament, to which a hook is attached, through the eye and secure it along the top and bottom of the hook by wrapping the entire shank. Such a two-hook rig is de rigeur for billfish flies.

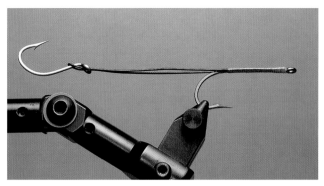

Secure a long loop of flexible wire or monofilament to the hook shank. You can interlock the loop through the hook eye and over the bend of the hook. If you don't need the stinger, the wire interferes little with the action of the fly, especially with flies with long hair or feathers, like Deceivers or Clousers. When called for, simply loop the wire or mono over the hook.

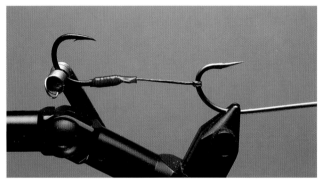

Although this is a commercial stinger hook, you can easily make a similar one with a loop of stout line tied to the hook eye. When needed, take the loop around the bend of the hook and loop the stinger hook though the loop.

Weed guards

All saltwater fishing doesn't take place in open, unobstructed water. Mangrove roots, grass beds, tree branches, rocks, oyster bars, or other structures—even floating grass torn free by currents—afford shelter for bait, while simultaneously providing cover for the predators stalking them. Fishing around such obstacles can be a nuisance. They can entangle and snare flies, or dull hook points. Nothing is more frustrating than retrieving a fly through a school of feeding stripers or past a tailing redfish and discovering that your fly was trailing a strand of eel grass. Although no solution guarantees 100 percent results, almost any fly can be rendered weedless or snag resistant to a great degree.

Due to their construction, flies like Bendbacks, Clouser Minnows, Jiggies, and most bonefish flies are less prone to hang up. They ride "upside down," with their hook bends up rather than down, making them snag resistant to varying degrees. However, an optional appendage can be added to almost any fly, either incorporated into the construction or added later to the finished fly, to make it even more weedless. Below are some common anti-snag devices, almost universally known as "weed guards." They afford varying degrees of protection to hook points, and each has its purpose and its champions. Keep in mind that the purpose of these deflectors is not to push the snag out of the way, but to push the fly aside. After all, you don't want to move the branch or rock, but rather make the fly glance off the obstruction, without hooking into or onto it. This means that you don't need a heavy or stiff guard in most cases. Try some of those shown and decide which are best for your needs. Anything that guards the hook point may cause some missed strikes, but I am convinced that this doesn't happen nearly as often as some fishermen imagine and that the benefits generally outweigh the occasional missed fish. For the sake of clarity, I've shown the various weed guards on bare hooks. With a few exceptions, it's a simple matter to adapt almost any one of the flies in this book.

Weed guards that can be added to the finished fly

You can make a single guard from a short piece of hard monofilament. As an alternative, you can fashion the same device from nylon coated braided wire or a piece of single strand, brown colored stainless wire. A weed guard of any of these materials can also be added to a hard body popper. Simply poke a hole in the body of the popping bug, dip the mono or wire in epoxy, and insert it into the bug body.

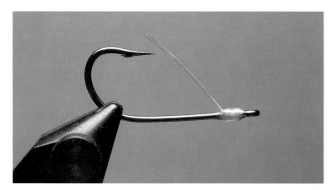

Single strand of monofilament: Flatten the end of the mono with flat-nosed pliers, tie it down to the shank or head of the fly, and trim to length.

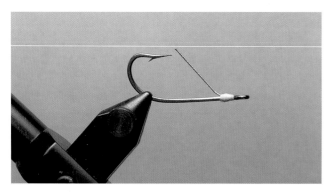

Nylon-coated, braided wire: Bend a slight kink in the wire with pliers and tie it in as with the monofilament. This makes a bit firmer guard than mono.

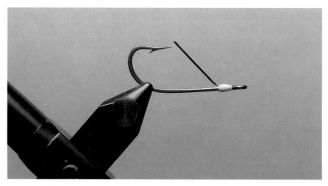

Single strand stainless wire: Bend this wire in the same way as the braided wire. It is stiffer and makes an even stronger guard.

A double guard gives additional protection and is recommended for larger flies.

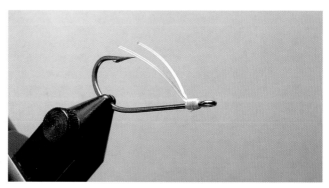

Double monofilament weed guard: Bend a piece of mono in half, flatten the point of the V, and fasten it behind the hook eye.

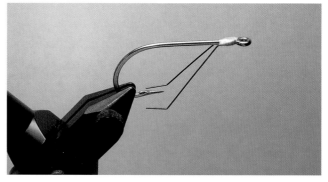

Double wire weed guard: As for the single guard, single strand wire, stainless wire can be used. Leave the legs of the wire extensions long, crimp them at an obtuse angle toward the rear, and snip off the excess.

This unusual deflector is highly effective. It causes a fly to hop over rocks, shells, and sticks without hanging up.

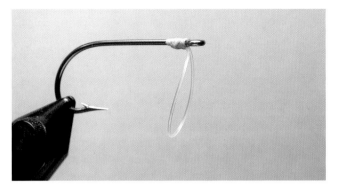

Crimp the ends of a short piece of mono, bend them 90 degrees, and attach so that the loop stands straight down, perpendicular to the hook shank.

A highly effective weed guard can be fashioned from bucktail coated with silicone and trimmed to length. It is especially adaptable to Bendback flies, due to the natural buoyancy of both the hair and the silicone, which keep the fly riding at the proper attitude. It works remarkably well and doesn't impede hooking fish.

Make a thin spike of bucktail.

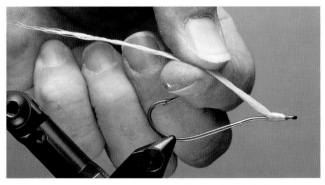

Coat the hair with clear silicone caulking. (See Siliclone tying instructions, page xx, for the use of silicone.)

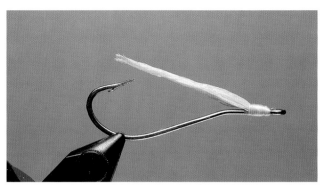

After the silicone cures, trim the hair to length, just beyond the hook point.

Weed guards that are incorporated into the construction of the fly

To create a single mono loop, the wire or mono is attached before the fly is begun and tied off after the fly is completed.

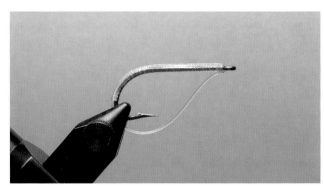

Tie in the mono along the hook shank and partway around the bend, and leave it protruding. When the fly is finished, bring the end forward and tie off at the head, beneath the shank.

Here's another effective and simple solution to the problem. I prefer this to the preceding.

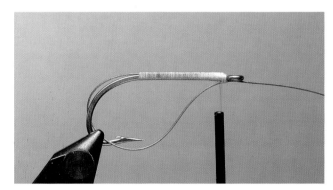

Attach the mono or wire to the side of the shank, take it around behind the bend, rest it against the inside of the hook bend, and tie it off at the head.

A double mono loop requires a little more work than the single loop, but is highly effective. For many purposes, this is my favorite. It can also be made of braided, coated wire.

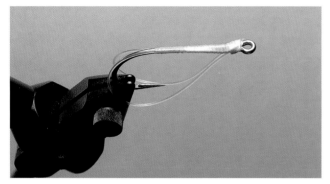

As with the single mono weed guard, the mono is attached before the fly is tied, and then wrapped off at the eye when the fly is finished.

A double weed guard from single strand wire is a device that requires a little more work than most but provides the maximum in weed guard protection in lily pads, mangroves, branches, and heavy grass. The wire takes on the look of two sled runners.

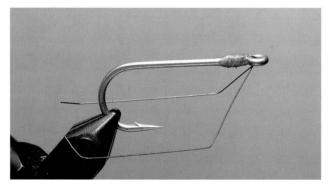

This guard can be added to the finished fly, or the wire can be tied at the start and folded forward out of the way, then bent back into position when the fly is completed. However, it can be a bit awkward working around the wire.

Here are four weed guards as they appear on finished flies.

CHAPTER 4

Baitfish Imitations

General purpose flies for striped bass, bluefish, weakfish, albacore, bonito, seatrout, redfish, snook, and other saltwater species

Joe Brooks's Platinum Blonde, a simple, generic bucktail, will catch almost anything that swims. Why, then, use anything else? Because fish can be particular, and until we know exactly what makes a fish strike a silly concoction of nylon or feathers or hair and tinsel, we must rely in large measure on trial and error. Yellow may tickle a fish's fancy one moment, blue a short while later. Change of light, the fish's visual receptors, water conditions, reflectance of materials: who knows what's responsible?

Our sport also demands that we make some reasonable, or even unreasonable, attempt to mimic natural forage. Realistic, impressionistic, and other such terms are applied to our attempts, but the fact remains that all our creations, even the seemingly most natural looking, are only vague likenesses of a gamefish's natural food. So, lots of baitfish imitations spring from the vises of tiers.

Anglers commonly categorize flies according to the target species sought: striped bass flies, tarpon flies, tuna flies, redfish flies, and so on. Fortunately, fish usually don't read the labels on their food. In a few cases, stereotypical fly designs have evolved that allow us to classify them as "bonefish flies" or "tarpon flies," and I've devoted separate chapters to those. Overall, however, it's difficult to classify flies based on the species sought. After all, striped bass don't know that they are eating "snook flies," and false albacore will greedily consume offerings intended for redfish. Many different species feed on sandeels, menhaden, silversides, or anchovies and will readily attack a bogus baitfish, regardless of the label or the specific name of the artificial. I've therefore chosen to group together many of the most popular and commonly used flies under the heading "General Purpose." They work for numerous inshore and offshore species, from seatrout to tuna. In the notes accompanying the individual flies, I specify what their originally-intended target species were, but this in no way should make you hesitate to use them for others. New fly designs are often born when tiers modify a fly designed for one species and adapt the original to different needs. Such is the case with Dan Blanton's Whistler, a successful west coast striped bass fly that has been converted to a successful tarpon fly. Lefty's Deceiver, designed to entice east coast stripers, has become a favorite for scores of species worldwide. And what fish won't eat a Clouser Deep Minnow, an idea conceived by a smallmouth bass guide on the Susquehanna River in Pennsylvania? And, so, I've included in this section fourteen designs that have proven themselves widely successful for numerous species. All are time-tested and proven winners. Choose the ones you like, practice tying them, and adapt them. Confidence in a particular fly is a major factor in its success.

BENDBACK

Materials

Hook: long shank, sizes 4-1/0
Thread: personal choice, to complement colors of the fly
Body: braided tinsel
Wing: bucktail; tinsel flash; saddle hackles; ostrich herl
Eyes: self-stick prismatic eyes

The Bendback is not a specific fly but a design that can be adapted to many patterns. The key is tying a hair or feather wing on a slightly offset long shank hook. This presents a streamlined profile; conceals the hook point; improves the fly's ability to shed snags and grass, since it rides hook up; and for many species, improves hooking. The buoyancy of hair and feathers, against the keeling effect of the hook's bend, keeps it riding "upside down." The following pattern is one example of a Bendback as I tie it. Possibilities are unlimited.

Tying procedures

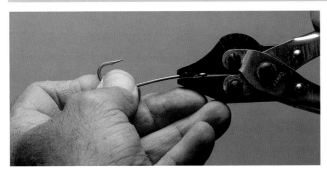

1. Grasp the front of the hook shank with pliers and bend down the shank with steady pressure to create a slight bend in the shank. Be especially careful not to make the bend too severe. I once clearly saw several redfish take my fly repeatedly without hooking up. When I removed some of the bend, I began hooking fish regularly.

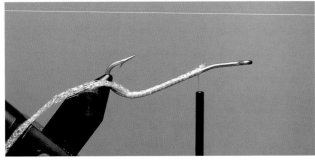

2. Attach the thread partway around the bend of the hook and tie in a piece of hollow braid along the shank of the hook and move the thread forward.

3. Wrap the braid around the hook up to the front bend in the shank; tie off and trim the end. Tie in a small bunch of white bucktail about twice the length of the hook shank.

4. Tie in a small bunch of your preferred flash (Flashabou, Krystal Flash, etc.) on top of the bucktail, and then add two white saddle hackles on each side of the wing, dull sides facing in. Feathers are shown here wetted for easier handling. When dry or under water, they will look fuller.

Additional tying notes and variations

5. Add contrasting saddle hackles to each side. I prefer grizzly hackles, either natural black and white or various dyed colors, because they give interesting barred effects.

This black pattern is a Florida favorite.

6. Tie in about eight or ten flues of ostrich herl over the top. This soft, breathing material adds life to the fly. Whip-finish and trim the thread; then add a prism eye to each side. It's now ready for a coat of head cement or a thin coat of epoxy.

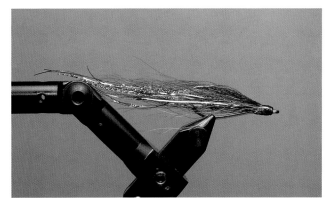

This sparse tie of two shades of bucktail and two different tinsels effectively imitates a bay anchovy.

A small Glass Minnow is tied on a Tiemco model 411, pre-shaped Bendback hook.

Lefty Kreh tied this all-black Keel Fly on a keel hook, a Bendback alternative. The bottom is coated with epoxy to increase durability when fished around rock jetties for striped bass. Lead wire can be added to any of these flies to carry them deeper in the water column.

This yellow version, also tied on a keel hook, is a favorite weakfish fly in New Jersey.

This creative fly, known as a Wide Side, was designed and tied by Captain Dave Chouinard of New Jersey. Successive bunches of Kinky Fiber are tied vertically up the shank of a circle hook. It makes an excellent imitation of a small menhaden, known locally as mossbunker, bunker, or pogie.

Fishing the Bendback

Redfish and snook hang close to cover, like grass beds, mangroves, or oyster bars, making the Bendback a logical choice.

BLONDE

Materials

Hook: standard or long shank, sizes 1-3/0
Thread: blue or white
Rear wing: bucktail, Flashabou
Top wing: bucktail
Body: flat silver Mylar tinsel

The legendary Joe Brooks, saltwater fishing pioneer, author of ten books, and world traveler before jaunts to now-fabled waters were fashionable, originated the Blonde series of flies. These simple, but basic, generic baitfish imitations are a good starting point for those getting into saltwater tying. They are as effective today as they were a half century ago. The blue and white Argentine Blonde (shown here) and others followed the original Platinum and Honey Blondes. This fly consists of not a tail and a wing, but two wings, the rear actually being slightly fuller than the top wing. When retrieved, water pushes the top wing back to give the fly the profile or silhouette of a baitfish. It's virtually foulproof.

Tying procedures

1. Attach the thread at the bend and tie in a wing of bucktail, fastening along the entire top of the hook shank. Make a smooth foundation with your thread.

2. Tie in several strands of Flashabou (blue and silver used here) at the bend over the bucktail. Move the thread forward, and attach a strand of medium or wide flat silver Mylar tinsel.

3. Wrap the tinsel to the rear and back again to the front; this will avoid gaps and thread showing through. Tie off the tinsel near the eye. As an option, you can coat the thread with clear head cement and allow it to dry before proceeding. Originally this step prevented metallic tinsels from tarnishing. Now it simply adds durability.

4. Using the same amount of bucktail as the rear wing, or less, add a top wing. Wrap a tapered head, whip-finish, and coat with cement.

Additional tying notes and variations

Building a good, smooth underbody of bucktail used for the rear wing, and wrapping the tinsel from front to rear and then forward again, will assure a neat fly. An oval tinsel body, or an oval rib over the flat tinsel, can provide extra flash. The original dressings for the Blonde series didn't call for flash in the wings. Now most tiers choose to add some flash to the wing, which definitely enhances the fly.

The all-white bucktail Platinum Blonde is one of the earliest striped bass flies and has stood the test of time.

The Honey Blonde is a basic alternative.

Fish sometimes prefer strong contrast, and the Integration Blonde provides it.

The orange and red Strawberry Blonde can be effective when fish key on bright colors.

The green and white Irish Blonde replicates the color of many common baitfish.

BUCKTAIL DECEIVER

Materials

Hook: Tiemco 911S or other long shank hook, 2/0-4/0
Thread: fine monofilament thread (or personal preference)
Body: bucktail
Flash: Angel Hair or other fine flash
Eyes: (optional) tabbed, self sticking prismatic eyes

Bob Popovics concocted this simple design, relying on the buoyancy, action, and natural taper of bucktail. Lefty's Deceiver inspired this fly, but the tying procedures, while simple, call for additional explanation and it merits a section of its own. It's an east coast staple. I am partial to this fly design because it allows me to readily control the profile, shape, and taper of my fly by manipulating one material. By varying the length, amount, and placement of the bucktail, I can make flies with rounded or flattened cross-sections, and produce imitations of small peanut bunker, cigar-like tinker mackerel, or large, slab-sided herring. Since the tips of the hairs provide most of the movement, the hundreds, or thousands, of individual hairs of lifelike, buoyant bucktail in a Bucktail Deceiver produce excellent swimming movement.

Tying procedures

1. Begin by attaching the thread at the bend and tie in a bunch of white bucktail, about two and a half times the hook length. Spread the hair evenly around the hook shank, not just on top.

3. After two or three segments of white hair, change colors, if you want to mimic the colors of a particular natural baitfish. Here two bunches of chartreuse have been added. Note that each bunch gets progressively a bit shorter and closer to the previous one.

2. Add a few strands of flash, advance the thread, and tie in a second bunch of hair, slightly shorter than the first, and more flash, as shown here. Note that the thread has moved forward up the hook shank.

4. Two bunches of blue have been added.

5. A final, shorter bunch of blue is added to give shape to the front of the fly. At this point, you can tie in tabbed prism eyes or simply whip-finish and cement the head.

Additional tying notes and variations

A dark topping and red tinsel throat can add realism and attraction.

Red and white is an ever-popular combination and works especially well on this fly.

This monstrous version, tied by Bob Popovics, covers an entire bucktail (notice pink underneath). It shows the lengths to which you can adapt certain flies. PHOTO BY BOB POPOVICS

Although it involves different tying techniques, Popovics's Hollow Fleye evolved from the Bucktail Deceiver. Using materials that don't flair, like the polar bear hair shown in this version, tie the hair sparsely, with the tip ends forward; then fold it back and tie the thread against the butts. The result is a large, long, but nearly weightless fly with a sparse silhouette and semitransparency.

Fishing the Bucktail Deceiver

Due to wind resistance, especially in larger versions, casting the Bucktail Deceiver requires a shorter, heavier leader and a longer, progressive casting stroke. In addition to striped bass, for which it was designed, in smaller sizes it has proved very effective for laid up tarpon, and its size and buoyancy make it devastating on pike in freshwater, where it suspends and fishes effectively over shallow weed beds.

Captain Dan Marini and the author display a nice bass that fell for a Bucktail Deceiver at Chatham, Massachusetts.

CLOUSER

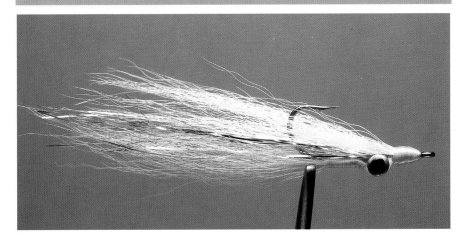

Materials required

Hook: standard length, size 8 to 2/0
Thread: personal preference, color to complement fly color
Wing: bucktail; Flashabou, Krystal Flash or other tinsel
Body: bucktail
Eyes: weighted dumbbell eyes, lead, tungsten or plated

Bob Clouser, the colorful smallmouth bass guide from Middletown, Pennsylvania, developed his Deep Minnow, almost universally known simply as a "Clouser" during the 1980s. Locally recognized as the "Commodore of the Susquehanna," he applied his vast understanding of bass fishing to producing a super-effective fly design, which Lefty Kreh subsequently christened the "Deep Minnow." Clouser also has become an outstanding saltwater angler, and his fly has found its way around the globe. Magazine publisher/editor John Randolph says of the Clouser Minnow, "It's the only fly I can never make a trip without, for any fish, fresh or salt water, anywhere in the world." It's probably safe to say that more different species of fish have been caught on Clouser Minnows than on any other fly. Like many great designs, the Clouser Deep Minnow is often copied and renamed. Bob Clouser's creative genius has also produced other great fly designs, including the Floating Minnow, Wounded Minnow, Crayfish, and Swimming Nymph.

Tying procedures

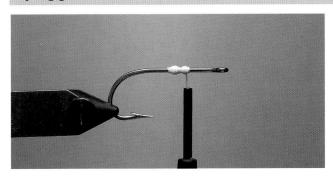

1. About one-third of the way back from the hook eye, form a small bump of thread, or two bumps, forming a saddle, as shown here.

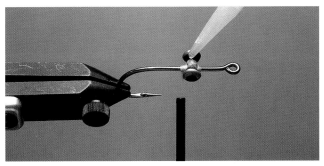

2. Attach the eyes by crisscrossing your thread and secure with a spot of CA cement. The eyes can be plain or pre-painted. Note: I prefer a silver iris with black pupil for most of my saltwater flies.

3. Attach a small bunch of bucktail, two to three times the hook length, just behind the hook eye. Don't wrap the thread back tight to the eyes.

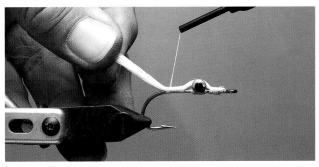

4. Bring the thread back to the rear of the eyes. Pull the hair over the bar of the eyes, back along the hook shank, and wrap down the hair on top of the hook shank.

5. Bring the thread forward to the front of the fly.

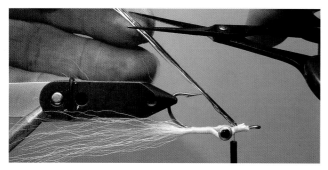

6. Turn the fly over or simply rotate the jaws 180 degrees if you have a rotating vise. Tie in six to twelve strands of flash or tinsel and trim to the length of the hair or slightly longer. Note: Cutting the ends of the tinsel to slightly different lengths gives better flash to the fly.

7. Tie in a contrasting bunch of bucktail hair behind the eye. Make this about twice as heavy as the first bunch, but the same length.

8. The basic Clouser, finished and ready to be whip-finished and coated with head cement. As an option, you can improve the fly's durability by coating the head with epoxy to protect it from the ravages of salt water and sand, at the same time increasing the weight of the head slightly.

Additional tying notes and variations

To turn over the Clouser and make it ride hook up, the lead dumbbell eyes are only a part of the story. The hair is far more important. The hair, being more buoyant, resists sinking, while the hook and the lead, being heavier, then sink first. If you tie a lot of hair on the "outside" of the hook, where the eyes are, it will sink with the hook down. See the "Tying Tips" section of the previous chapter for painting Clouser eyes. Tie your Clousers with different size/weight eyes for different situations. *Note:* See the following section for instructions on tying the Half and Half, a Clouser/Deceiver hybrid.

The diminutive Bonefish Clouser is deadly for bonefish, in tan, chartreuse, or pink over white.

I like to tie all the hair on the inside of the hook for a more streamlined minnow effect.

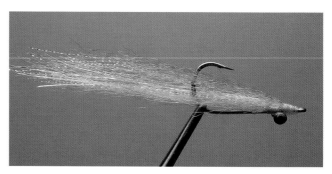

Renowned North Carolina angler Tom Earnheart dubbed this sparse, short version using Super Hair or Ultra Hair the Alba-Clouser, because of its success with false albacore. I tie it in blue/white and green/white, but am partial to this tan/white version when fish are feeding on bay anchovies.

The rabbit-tail Clouser possesses great action. Cut a small slit in the rabbit hide, slip it over the hook, and fasten it down just behind the eyes.

Fishing the Clouser

While Clouser himself usually attaches the fly with a fixed knot, like a clinch, I prefer the nonslip mono loop knot, which gives a little more freedom and action to the fly, especially for small versions like the alba-Clouser, Foxee Clouser, and Bonefish fly. Try both and decide for yourself. This fly can be fished many ways. Try making long sweeping strips, then pause, let the fly drop, and repeat. Bob Clouser has a deadly technique that involves making a long sweep with the line hand, finishing with a sharp flick on the line with his thumb, which makes the fly suddenly dart forward before falling. Occasionally, a series of short, quick strips that work the fly in a rapid, up-and-down pulsing manner will get a response.

Any fish that can be taken on a fly will eat a Clouser. Lance Erwin caught this mutton snapper in the Baja surf.

John Zajano unhooks a large weakfish caught on a pink and white Clouser Minnow in Barnegat Bay, New Jersey.

Bob Clouser took this hefty false albacore on his namesake fly at Cape Lookout, North Carolina.

DECEIVER

Materials

Hook: standard or slightly longer shank, typically in sizes 2-4/0
Thread: Danville flat waxed thread, or your preference
Wing: saddle hackles, not too thin
Flash: Flashabou or other flash
Body: flat or braided tinsel
Collar: bucktail
Throat: red Flashabou
Eyes: stick on prism eyes with tabs

Lefty's Deceiver is the quintessential saltwater fly. Developed in the 1960s by Lefty Kreh, the ultimate fly-fishing guru, it was designed to answer the most important requirements of saltwater flies: general baitfish profile, simple design, castability, non-fouling materials, and variation for specific needs. Perhaps no fly has inspired more adaptations and variations, including the Bucktail Deceiver described earlier. Nearly any species can be caught on a Deceiver. If you were limited to one fly, this would be it.

Tying procedures

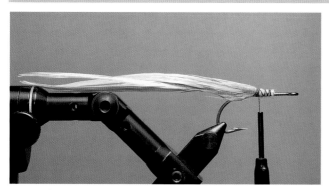

1. Prepare 6–8 saddle hackles, 3–4 for each wing. Moisten the feathers as shown and tie in the wings (individually or both at once) at the bend. You can face the dull sides inward or turn them outward, as shown here, for more action.

3. Attach a piece of tinsel or hollow braid at the rear of the shank, wrap forward, and tie it off.

4. Tie in a small bunch of bucktail low on each side of the hook.

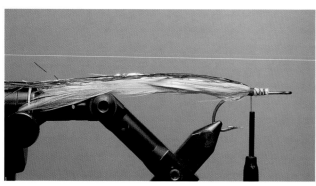

2. Tie in several strands of flash over the wings.

5. Add a third, longer bunch on top of the hook shank.

Long, dark topping helps give this fly a wider profile to imitate a herring or menhaden.

6. Tie in a short bunch of red Flashabou for the throat, attach an eye to each side, and whip-finish.

This Squid Deceiver, tied by Captain Dan Marini, who guides out of Chatham, Massachusetts, is a staple when squid arrive in great numbers around Cape Cod in June and the large bass gorge on them. He counts on the marabou fluff at the base of the hackles to give the fly a full appearance, and uses shorter bucktail collar than shown above on the standard fly. Note too the large prism eye placed well back on the outer feather.

Additional tying tips and variations

Lefty's Deceiver can be tied from two inches to a foot in length. You can employ up to a dozen hackles in the wing, depending on fly size. Make sure the bucktail collar extends well past the bend of the hook, to maintain the baitfish profile. Use bucktail from near the tip of the deer tail; that near base is too buoyant, won't sink well, and splays when cinched down with thread. For smaller flies, and when sinking is important, calf tail works well in place of bucktail. It has limited length, but it is not nearly as buoyant and sinks more readily.

A wider, more compact fly is achieved simply by making the hackle and feathers the same length.

Contrasting barred grizzly hackles are often used on Lefty's Deceiver.

Fishing the Lefty's Deceiver

Mix your retrieves, using long sweeping strips, with pauses, or quick, short, pulls in rapid succession.

The flats of Barnegat Bay, New Jersey, yielded up this 11-pound weakfish that fell for the author's red and orange Deceiver.

Bill Schotta offers proof that Lefty's Deceiver still takes bass in the Chesapeake Bay, where Lefty first developed it years ago.

This 35-pound jack crevalle grabbed the author's large white Lefty's Deceiver at Cape Lookout, North Carolina.

This bass fell for a large Lefty's Deceiver, which effectively imitated the menhaden on which it was feeding in Sandy Hook Bay, New Jersey.

HALF AND HALF

Materials

Hook:	standard length sizes 2-4/0
Thread:	personal preference
Wing:	saddle hackles
Collar:	bucktail
Body:	tying thread over bucktail
Top Wing:	bucktail; Flashabou, Krystal Flash, or other tinsel
Eyes:	lead, tungsten, or plated dumbbell eyes

The Half and Half came from the vise of Bob Clouser and combines the two most popular saltwater flies in the world: Clouser's own Deep Minnow and Lefty Kreh's Deceiver. While this fly could have been included as a simple variation of the Clouser Deep Minnow, I feel it is so important that it deserves separate and fuller treatment. Review the tying procedures for Lefty's Deceiver and the Clouser Deep Minnow before tying this fly.

Tying procedures

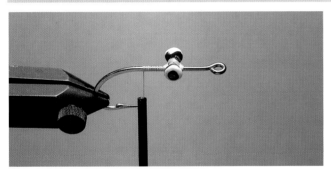

1. Form a bump or saddle on the shank about one third of the way back from the hook eye, as shown in the description for the Clouser Deep Minnow. Attach the eyes by crisscrossing the thread and secure with a drop or two of CA cement.

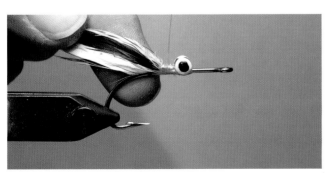

2. Make two sets of saddle hackles, with three feathers in each. Make them about three times the hook length. Wet them to make them more manageable. Place the dull sides together and tie both sets on top of the hook shank. Here the outside hackles are of red dyed grizzly for contrast and color. Secure with extra wraps of thread.

3. Tie in a fairly large bunch of bucktail near the bend, where the hackles were tied in. The hair should be about twice the hook length. As you tie down the bucktail, spread it evenly around the hook shank. This completes what is effectively the Deceiver part of the Half and Half.

4. Wrap the thread forward. Tie in a bunch of bucktail, not too heavy, behind the hook eye, as for the Clouser Deep Minnow.

5. Take the thread behind the eyes, pull the hair back over the bar of the eyes, and fasten it down securely.

6. Take the thread forward toward the hook eye. Turn the hook upside down or rotate the vise 180 degrees. Add six to twelve strands of Flashabou, Krystal Flash, or other similar tinsel tied in ahead of the eyes.

7. Tie in a bunch of bucktail, the same or a contrasting color, and the same length as the collar of the Deceiver half of the fly. Whip-finish and coat the head with head cement. Once again, coating the head with clear 5-minute epoxy is an option for durability and additional weight.

Additional tying notes and variations

Chartreuse, alone or in combination with other colors, makes a deadly combination.

You can tie the Half and Half in endless color combinations to match bait colors and the whims of fish.

Fishing the Half and Half

Use the same retrieves and techniques as described for the Clouser Deep Minnow. Take a good assortment of Half and Halfs, along with regular Clousers and Deceivers on destination trips, when you aren't sure what critters you'll be catching.

Captain Dan Marini hoists a nice bass Lefty Kreh took on a long white Half and Half in Chatham, Massachusetts.

HI-TIE

Materials

Hook: regular or long shank, sizes 2-3/0
Thread: flat waxed
Wing: bucktail
Flash: Flashabou

This is another universal baitfish imitation that is easy to create and doesn't foul. Like the Seaducer and Blonde flies, it relies essentially on one material, with a touch of flash. The fly's shape comes from tying successive bunches of bucktail along the top of the hook up the shank. Though different in shape, like the Bucktail Deceiver, it derives life in the water from the breathing action of hundreds of moving, tapered hairs. It is a traditional freshwater bass fly concept that has been adapted to salt water. Mark Sosin popularized the Hi-Tie years ago as the Blockbuster, and many anglers still know it by that name. The Hi-Tie's wide profile makes a satisfactory imitation of a menhaden, herring, or other wide baitfish. It's a great fly for learning to handle and control bucktail.

Tying procedures

1. Attach the thread at the bend, above the barb, and tie in a bunch of bucktail on top of the hook shank off the bend. Make the hair one and a half to two times the hook length, but not too heavy.

2. Add a few strands of Flashabou or other tinsel. Keep it sparse.

3. Tie in another bunch of bucktail, topped with more flash, keeping it all on top of the hook shank. Use hair the same length as the first bunch.

4. Repeat step 3, tying in additional hair and tinsel until the hook shank is full. Making the last bunch sparser and shorter will give the fly a better silhouette. After tying off the thread, lacquer the head and the bottom of the thread wraps along the shank with head cement.

Additional tying tips and variations

As an alternative to using a long shank hook, a hook like the Varivas 994S hook shown in this sample has a longer straight shank than most standard length shanks, due to its peculiar bend. Use hair from the top portion of the bucktail. That near the base has less action, tends to flare more and, being more buoyant, keeps the fly from sinking readily.

You can mix the colors or add bars or stripes to the hair with a permanent marker. The pink Hi-Tie is a traditional favorite.

The Hi-Tie readily lends itself to blending various colors, as this mix of white, green, and blue indicates.

JIGGY

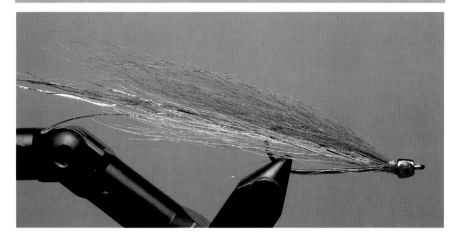

Materials

Hook: long shank, sizes 4-3/0
Thread: clear mono (or color to match top bucktail)
Wing: bucktail
Head: Jiggy head (or brass or tungsten cone head)
Eyes: prismatic (or Krystal Jiggy head with pre-installed eyes)

Bob Popovics conceived this fly to produce a jigging motion, in the tradition of the Clouser Deep Minnow, but with some significant variations. First, the cone head or Jiggy head places the weight farther forward. This feature, along with a generally sparser dressing, causes the fly to drop sharply between pulls when retrieving. Overall, the Jiggy gives a more streamlined minnow shape, making it especially suited for imitating thin baits like sandeels and spearing. Also, the head cannot break off, as lead dumbbell eyes will sometimes do when they strike a hard object.

Tying procedures

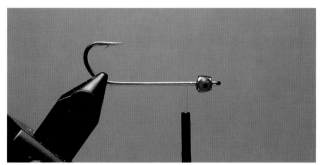

1. Slide a Jiggy head (or cone head) to the hook eye. For more weight and sharper jigging action, you can wrap a few turns of lead wire around the shank and push it inside the head (*See* "Tying Tips" section in chapter 3).

3. On top of the hair, add several strands of Flashabou or other flash, extending beyond the ends of the hair.

2. With the hook in the vise "upside down," and the thread attached just behind the head, tie in a sparse bunch of bucktail hair, about twice the hook length. Important: keep all material on top of the shank. This is the most important factor in assuring that the fly rides hook up. If it is spread around the hook shank, the fly will likely ride with the hook down.

4. Add a sparse bunch of a contrasting color bucktail, about the same length as the first. Tie off the thread and coat the wraps with head cement.

Additional tying notes and variations

You can make the hook slightly Bendback style (*See* "Bendback" tying instructions for additional assurance that it rides hook up). Other options include using a rabbit strip instead of or in combination with the hair wing and including an extension of monofilament from the bend, onto which are tied several saddle hackles, to make an extra long fly.

An olive and white cone head Jiggy is my favorite sandeel imitation. When using a cone head, place the eyes over the wraps and coat with a thin coat of clear 5-minute epoxy. If you use clear mono tying thread, you can secure the eyes with it before applying the epoxy. The thread becomes virtually invisible when you coat it.

This bucktail version has a metal bead plus a cone for added weight.

The Keel Eel, a simple sandeel or elver imitation, can be fashioned from Super Hair or Ultra Hair on a long shank, Bendback hook.

Fishing the Jiggy

Attach the fly to your leader with a loop knot for better action, and use the same retrieves as described for the Clouser Deep Minnow.

A 40-inch striped bass caught on a six-inch olive and white bucktail Jiggy, which effectively imitated sandeels on which the bass were feeding in Chatham, Massachusetts.

Jiggies in blue, green, and tan over white effectively imitate the smaller baits on which false albacore and bonito commonly feed. They are among the top albie flies used on the east coast.

SEADUCER

Materials

Hook: standard length, sizes 1/0-3/0
Thread: 3/0 or flat waxed
Wings: large cock neck hackles or saddle hackles
Hackle: same as wing

Well-known Florida fishing writer/angler Chico Fernández popularized this fly, although some form of it has been around for generations. It's similar to the Homer Rhode Tarpon Fly, except that the entire hook shank is closely palmered (spiraled) with hackle.

Tying procedures

1. Attach the thread at the bend and tie in two or three wide neck hackles or saddle hackles on each side, curving outward. Wetting them first will make them easier to handle and control.

3. Grasp the tips of the hackles, together or one at a time, and wrap them around the shank with the turns close together. Tie down and trim the ends of the hackles.

2. Tie in two hackles, the same color as the wings, by the butts.

4. Tie in two contrasting colored hackles by the butts.

5. Wrap them as you did the first two and tie off.

6. Attach and wrap two more hackles of the first color. Tie off the thread with a whip-finish, and coat the head with head cement.

Additional tying notes and variations

This simple fly is easy to tie and begs for some flash. Most often tiers add Flashabou or Krystal Flash at the rear, after attaching the wings. Tiers often use saddle hackles in place of long neck hackles. The flared neck hackles have stiffer stems and give more pulsating or kicking action to the fly. Saddle hackles give more swimming motion and length. You can also begin by tying in a slender bunch of bucktail at the bend to help separate the wing feathers. Some tie the Seaducer all one color, except for the final hackle at the front, but the banded effect is more common.

A top redfish and snook fly is this yellow and grizzly version, tied with shorter wings and the collar hackles pulled back and held in place by wrapping the thread against the base of the hackles.

I've had excellent results with the black and purple version, also with saddle hackles dull sides inward, when night fishing for striped bass.

The red and white hackle fly is the most popular version of this fly in fresh and salt water. I prefer it with saddle hackle wings, dull sides facing inward.

Fishing the Seaducer

The Seaducer is a staple in the arsenal of redfish anglers in Florida and along the Gulf coast. Yellow and grizzly or green and grizzly are the most popular colors.

SILICLONE

Materials

Hook: standard or short shank, either with wide gap (or use one size larger than regular), sizes 1/0-4/0
Thread: clear mono thread (or personal preference)
Wing: bucktail (flash optional)
Body: sheep fleece or ram's wool; silicone caulking (GE or DAP both are excellent; Kodak Photo-Flo solution from camera shops, although increasingly hard to locate)
Eyes: self-sticking prismatic eyes

Like epoxy, acrylic, hot glue, and other nontraditional substances, silicone offers potential previously unavailable to fly tiers. Using silicone as described here will enable you to make bodies of various shapes—wide, flat, or round. The pattern shown is the mullet Siliclone. Black and yellow are the other two colors I favor highly. It has great buoyancy for near-surface fishing.

Tying procedures

1. Attach the thread at the bend and tie in a bunch of white bucktail, not too heavy but quite long. Here a short hook is being used, so that the hair is about three times the hook's length. Distribute the material around the hook shank.

2. Cut a bunch of sheep fleece from the hide and push it over the hook shank from the eye, with the tips facing the rear so that it forms a veil around the hook. Fasten it down tightly with several wraps of thread. Work the thread forward of the fleece and make several more wraps, before adding more fleece.

3. Cut another bunch of fleece from the hide but cut off the tips before attaching it to the hook as with the first bunch. Tie it down tightly and work the thread forward again.

4. Repeat step three until the shank is filled with fleece. As you work forward, be sure to pack and compress each bunch as tightly as possible before tying in the next one.

52 ▪ BAITFISH IMITATIONS

5. Whip-finish and trim the thread. With your scissors, begin trimming the excess fleece. Make the first cuts on the bottom, making sure to leave sufficient hook gap clearance.

6. Continue trimming the top and sides until you get the shape desired. It is crucial that the fleece be smooth and even, with no hollow spots or high points. Silicone will not conceal irregularities, but will actually accentuate them.

7. This top view reveals a wide, flat head, much like that of a natural white mullet. The Siliclone gives you the opportunity to make baitfish imitations in various shapes and sizes.

8. Using your fingertip in a light feathering motion, brush a smooth and even layer of silicone over the entire fleece body. Coat the area around the hook eye well, to help keep water out.

9. After about 10 minutes, apply prism eyes. Follow this with a second layer of silicone. Periodically dip your fingertip in Photo-Flo solution. This will smooth out the silicone, as if you were running your finger over soft butter. Let the fly cure, preferably overnight.

Additional tying notes and variations

Use fleece that is somewhat wiry and springy to the touch. Soft, plush materials like egg yarn don't work so well. If the hair is knotted and unruly, brush it with a wire dog comb to get out the tangles before clipping it from the skin. When you trim the tied-in fleece, rather than having the hook dead center, keep about two thirds of the material above the shank and one third below. As an alternative to Photo-Flo, you can use clear dishwashing detergent, vegetable oil, or glycerin (available from drug stores) as a wetting agent. Packing the fleece tightly and carefully trimming it evenly to get a smooth, velour feel will result in a smooth finish when the silicone is applied.

The Siliclone, especially when tied Bendback style and with a bucktail weed guard added, is almost completely weedless (See "Tying Tips" in chapter 3).

A lipped version, the Pop Lips, duplicates the side-to-side swimming motion of hard-bodied plugs. The lip is made of trimmed fleece coated with silicone. This particular sample, tied by Bob Popovics, also features a "stinger" tail hook.

A Siliclone worked slowly through inlet waters is the author's favorite night fly. This bass was taken on Martha's Vineyard, Massachusetts, at night.

This large Siliclone, tied with a wide rabbit strip wing, provides great swimming motion when the fish want more action.

Fishing the Siliclone

Connect the Siliclone with a loop knot for freer action. It works especially well when fished exceptionally slowly at night, swimming just below the surf and causing a slight wake. Use a hand-over-hand retrieve to keep it moving steadily without interruption. The object is to make the fly push water ahead of it, sending out vibration waves. The air pockets in the fleece help the fly to float. If it absorbs water and begins to sink, simply squeeze out the water. Conversely, if you want it to sink, to fish over submerged structure on a sinking line, squeeze it tightly, then hold it under water and release the pressure—the fly will suck in water. The peculiar splat that a Siliclone makes when landing appears to attract fish. It's deadly on striped bass along on the Atlantic coast and has been an effective snook fly in the Everglades.

The Siliclone works especially well, when smacked down in white water. This bass grabbed Lance Erwin's white Pop Lips Siliclone in the turbulence along the rocks of a New Jersey jetty.

SLAB SIDE

Materials

Hook: standard length, sizes 1/0-4/0
Thread: personal choice
Wing/tail: mixed bucktail and flash
Body/head: spun and trimmed deer body hair
Eyes: glass or plastic doll eyes

New England fly fishing celebrity Lou Tabory came up with this design. It simulates a wounded baitfish, generally a herring or menhaden. The blend of hair, flash, and peacock give the Slab Side a natural appeal. The fly is buoyant due to the amount of deer hair used. It fishes well on a sinking line, which pulls it down while the fly struggles to float upward. I like this fly because it behaves a little differently from most other flies, and I'm always looking for options in my fishing.

Tying procedures

1. Starting with your thread attached to the shank at the bend, just above the hook point, tie in a medium bunch of bucktail. Note that all wing materials are kept on top of the hook shank to give the finished fly a wide profile.

3. Tie in a medium bunch of pink bucktail. Try to keep the materials tightly wrapped and close. Note that the thread is barely halfway up the shank at this point.

2. On top of the bucktail, tie in a small bunch of white synthetic hair. Mixing the more translucent synthetic with the more opaque natural hair adds a little iridescence and sparkle. Add some flash, in this case pearl Flashabou.

4. Add a last bunch of bucktail, this time light gray.

5. Tie in about 12 strands of peacock herl on top of the bucktail, and trim off the butts. This represents the dark back of the baitfish.

6. Tie in a bunch of deer body hair next. Wrap it with two turns of thread, and pull it down to the shank. Continue pulling down on the thread as you release your grip on the hair and it will spin and distribute itself around the hook shank.

7. Pack the hair tightly with your fingers or a hair pusher, as described in the "Tying Tips" section of chapter 3. Now work the thread forward in front of the hair, ready to tie in the next bunch.

8. Tie in additional hair until the shank is filled, packing each bunch tightly up against the previous one. Then whip-finish, cut the thread, and secure it with some head cement.

9. Trim the head with your scissors, leaving it large but flat on the sides. All that remains is to place a plastic doll eye on each side, securing them with a spot of CA glue.

10. Note the flattened sides of the deer hair head in this top view.

Additional tying notes and variations

Lead or tungsten eyes can be used if you desire a weighted fly. Fasten the eyes to the shank after the first bunch of deer body hair is tied in, so that the eyes are about one-third the shank length back from the hook eye. Optionally, you can give the bottom of the finished fly a light coat of epoxy or CA cement to balance the fly. I actually like it with some fine lead wire tied around the shank before commencing the dressing.

SURF CANDY

Materials

Hook: standard or short shank salt water hook, commonly size 4 through 1/0
Thread: fine monofilament thread
Wing: Super Hair or Ultra Hair; Flashabou
Body: "Tuffleye" clear acrylic (manufactured by Wet-A-Hook Technologies)
Eyes: self-sticking prismatic eyes

Bob Popovics's Surf Candy evolved from years of experimenting with epoxy. The original Pop Fleye, it was tied with durability in mind—more than 30 bluefish have been caught on a single Candy. Surf Candies, replicating small saltwater baitfish like silversides (spearing), bay anchovies, and sandeels, ushered in a whole range of new flies, and are among the most copied flies anywhere. In addition to being a staple for striped bass and bluefish anglers, one New England captain claims that they "revolutionized albacore and bonito fishing." It's hard to find a saltwater angler whose fly box doesn't sport some version of the original Surf Candy, regardless of the name. Using only a few materials and basic procedures, Popovics's intention was to sculpt a fish body with epoxy. You can modify Surf Candies in countless ways, to take advantage of translucency, realism, body form, flash, color, and weight. You can add feathers, Mylar strips, or bead heads or employ Craft Fur or rabbit strips. To increase or decrease the fly's weight or change the color, tiers have added tungsten powder, micro-balloons, or food coloring.

Bob tied the basic fly shown here. Note that he ties this one without epoxy, using clear acrylic instead, just the latest in an endless series of innovations. It is lighter than epoxy, but it is clearer, easier to use, doesn't require mixing, won't yellow, and is virtually indestructible.

Tying procedures

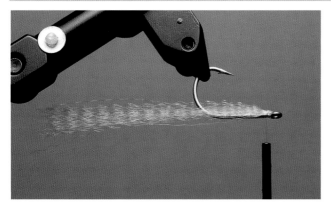

1. After attaching the thread at the front of the shank, tie in a bunch of white or polar bear-colored Super Hair behind the hook eye on the bottom of the hook shank. Before you attach it, trim the butt ends square, not tapered, and fasten it down only in the front of the fly, behind the eye. A rotating vise helps with this step. Otherwise, place the hook upside down in the jaws.

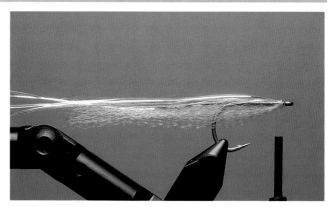

2. Turn the hook over and add some flash, like pearl Flashabou, on the top of the shank.

3. Prepare a second bunch of hair as above, and attach it to the top of the fly. This is normally in a contrasting color. Tan is used here. Whip-finish and tie off the thread.

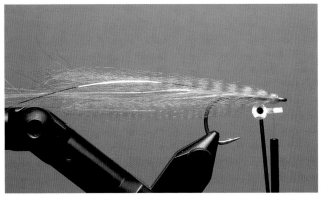

4. Attach a prismatic eye with tab on each side of the head.

5. Squeeze acrylic "Core" coat over the hair and cover the head wraps. Work it into the hairs. There is no need to rush, since, unlike epoxy, acrylic doesn't run and won't begin to harden until exposed to a blue light.

6. Don't apply acrylic beyond the bend of the hook. Add material or remove some with a bodkin until you are satisfied with the body shape.

7. Following the instructions that accompany the product, expose the acrylic to a visible spectrum blue light (available from Wet-A-Hook), being careful not to shine it directly into your eyes. Hold the light close for a few seconds and move it over the entire body.

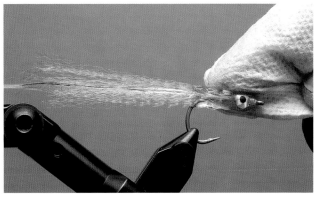

8. The acrylic will have a faintly tacky feel. The purpose is to promote better bonding to the optional top or "finish" coat of acrylic, if you decide to use that for a slicker finish. Otherwise, simply rub the acrylic with a paper towel moistened with rubbing alcohol. This gives a subtle, dull, but smooth finish.

Additional tying tips and variations

See "Materials" section of chapter 2 for more information about acrylic. Trim the hair to length with a taper, and let the tinsel extend a little beyond the hair for a "flashtail" effect. You can use a red marking pen to add gill slashes to Candies.

This large, bead head version, known as a Deep Candy, is coated with epoxy, instead of acrylic.

A Schoolie: Multiple small Surf Candies are tied on a single hook to simulate a small "bait ball."

Not only do acrylic flies not yellow, they actually get clearer when exposed to sunlight. PHOTO BY BOB POPOVICS

Epoxy makes flies heavier but will yellow when exposed to sunlight over a period of time. For this reason, I recommend tying the flies in advance but not adding the epoxy until shortly before fishing them. PHOTO BY BOB POPOVICS

Fishing Surf Candies

For its realism and durability, a Surf Candy is the ideal choice when bluefish drive bay anchovies or silversides to the beach and action heats up.

Few flies can withstand the savage teeth and jaws of bluefish as a Surf Candy can.

WHISTLER

Materials

Hook: standard length, sizes 2-4/0
Thread: red 3/0
Wing: bucktail
Body: chenille
Hackle: saddle hackle
Eye: heavy bead chain

California fly-fishing guru Dan Blanton designed the Whistler primarily for striped bass in the San Francisco Bay area. Tiers interpret the fly's basic theme in a variety of ways, and Whistlers are effective for a wide range of game fish. It certainly belongs in any list of top saltwater flies. Whistlers are so commonly used for tarpon in Central America that I included complete step-by-step dressing for a Tarpon Whistler in the chapter on "Tarpon Flies," rather than simply listing it here as a variation of the basic design.

Tying procedures

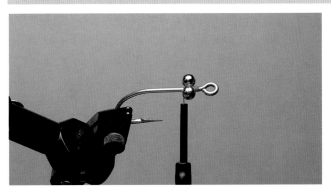

1. Attach the thread a short distance behind the hook eye, and fasten large bead chain eyes with crisscrossed thread. Secure the wraps with head cement.

3. Fold back the Flashabou and tie it down, leaving the ends quite long. Tie in a bunch of bucktail at the bend, spreading the hair evenly around the hook shank.

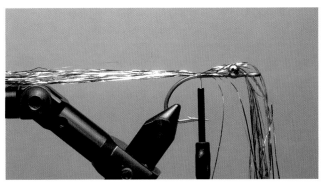

2. Take the thread to the rear of the shank, and tie in a generous bunch of silver Flashabou in the middle of the strands.

4. Wind the thread back slightly to keep the bucktail from spreading.

5. Scrape about ¼ inch of fluff from the end of a piece of chenille, exposing the core, and attach the thread core to the hook; then take the thread forward.

6. Wrap the chenille around the shank and tie it off at about the middle of the shank.

7. Tie in two saddle hackles by their butts, dull sides to the rear.

8. Wind the hackles several turns, tie off, and trim the excess. Cut the Flashabou tail so that it extends about ⅜ inch past the end of the bucktail. Whip-finish and coat the head with cement.

Additional tying notes and variations

This sample was tied by Dan Blanton. He often dresses his flies heavily for large bass and now prefers the Eagle Claw 413 60-degree jig hook for his Whistlers and other flies. The "flashtail" tinsel extension is a trademark of Blanton's flies.

The overall impression of Blanton's Whistler is unmistakable, despite various color combinations.

WOBBLER

Materials

Hook: standard length, sizes 2-3/0
Thread: flat waxed nylon
Tail: Krystal Flash or Accent; bucktail, calf tail, or synthetic hair
Body: braided Mylar tubing with the core removed; 5-minute epoxy
Eyes: black bead chain

Recalling the look and action of classic spinning and casting spoons, the Wobbler is the brainchild of Floridian Jon Cave, a versatile and talented angler, tier, guide, and instructor. Cave's Wobbler has added a dimension to saltwater fly tying that few others have achieved. I've tried several other spoon flies, but none works like Cave's fly. Instead of wobbling, hook up, with a side-to-side swinging motion, most either simply drag through the water with little action, or spin and rotate, twisting the leader. Originally intended as a shrimp imitation for redfish, the Wobbler is effective for many species. The instructions here vary slightly from Cave's original. He puts epoxy inside the braid, while this version, concocted in collaboration with my tying "model" John Zajano, is coated only on the outside. The body doesn't need much curve, but the weight at the bend is critical for proper action.

Tying procedures

1. Start with the hook in the vise, eye pointing downward, and attach the thread one-third of the way around the bend. Tie in a few short strands of gold Krystal Flash and a short tuft of bucktail, calf tail, or Kinky Fiber on top of this as shown.

3. Pull out the cotton core and iron the tubing flat. This largely eliminates the need for putting epoxy inside to hold the shape.

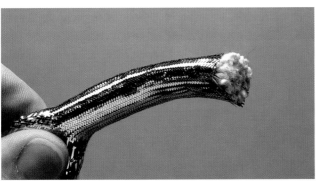

2. Cut a few inches of braided Mylar tubing. Use 1/4 inch or 3/8 inch as required by the hook size.

4. Slip the tubing over the hook shank, and tie down the tubing at the bend. Whip-finish and tie off the thread.

5. Reattach the thread at the eye and tie down the tubing tightly at that point. You may have to fiddle with the tubing to get it even on each side of the hook before securing it tightly. Trim the excess tubing, whip-finish, and cut the thread.

6. Reattach the thread at the bend and add a pair of large (or extra large for the biggest flies) black bead chain eyes. Tie off the thread for the final time.

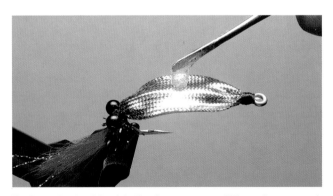

7. Coat the braid with 5-minute epoxy. When the epoxy is nearly completely hard, squeeze the braid to get the final shape. Dip your fingers in Photo-Flo solution, dishwashing detergent, or vegetable oil to keep them from sticking to the epoxy. At any rate, the epoxy will be only slightly tacky at this point. Add a second light coat when the first is completely cured.

Additional tying notes and variations

The instructions given above are for the standard C-Wobbler. Cave also ties an S-Wobbler on a bendback style hook (see Bendback flies), which has a slightly quicker action. Also, his Lil' Wobbler uses folded glitter tape, glued to the shank and coated with epoxy, instead of tubing over the shank. Vary the color of the tubing and the hair to suit your whim. The action is what counts.

Since redfish and snook are often close to grass beds or oyster bars, a Wobbler tied on a weedless hook, like this version from Umpqua Feather Merchants, is a good idea.

Fishing Cave's Wobbler

Jeff Schneider displays a large Louisiana redfish taken on a Wobbler.

CHAPTER 5

Crabs and Shrimp, Eels and Worms

The vast majority of saltwater flies imitate baitfish. Most of the flies presented thus far were designed to imitate spearing, anchovies, sandeels, herring, menhaden, pinfish, and other typical schooling finfish. I've devoted this chapter to "non-baitfish" flies to draw attention to them. They are too often overlooked and too often important enough not to be. While some bonefish flies are designed specifically to simulate shrimp, some shrimp flies are universal or crossover patterns, less specifically species oriented. And crabs are standard fare for permit, redfish, weakfish, seatrout, striped bass, snappers, and scores of other species. Eels are important to striped bass fishing, and striper and tarpon fishermen have, relatively recently, become aware of the importance of various worms in the diet of their target species. I've devoted this chapter to a few reliable imitations of each of these baits.

Not only do decapods (crabs, shrimp, lobsters) and annelids (worms) look different from typical baitfish, they behave differently. This means that as a tier you must learn to work with different fly designs, and as an angler you must learn to employ other fishing techniques or retrieves. Tying and fishing these flies will make you a well-rounded tier—and a more successful angler. Cephalopods, including squids, constitute yet another group of common baits. However, I have opted to forgo separate treatment of these, as well as some others, first because I had to draw a limit to the topics to include and second because some of the flies discussed elsewhere have served well as squid imitations. Note especially Dan Marini's Squid Deceiver, shown in the section on Lefty's Deceiver. Bendback, Bucktail Deceivers, and Whistler flies can also be modified slightly to make reasonable counterparts of squid.

CHERNOBYL CRAB

Materials

Hook: standard length, size 2 or 4
Thread: white
Tail: Krystal Flash; white calf tail, furnace (brown with black center) hackles
Body: deer body hair; saddle hackle
Eyes: painted lead dumbbell eyes

Tim Borski, the famous artist/angler out of Florida, designed this fly. It's one of the best crab patterns to come down the pike. If you can't take a bonefish, permit, striped bass, or redfish on this fly or the Del Brown crab that follows, they aren't eating crabs. Like Borski's Swimming Shrimp, this fly has the buggy look tiers and anglers generally favor. It enters the water with a decided and unique splat and looks totally alive when crept across the bottom.

Tying procedures

1. Starting with the thread attached at the bend, tie in a small bunch of Krystal Flash, cocked slightly down. On top of this, tie in a small bunch of calf tail.

3. Stroke a brown saddle hackle to make the fibers stand out from the stem. Trim the tip of the hackle off, and clip the flues from each side of the stem. Leaving just short stubble will provide secure gripping for the thread.

2. Attach two short furnace-colored hackles, curved outward, on each side of the tail.

4. Cut a small bunch of deer body hair from the hide and position the hair over the hook shank with two wraps of thread. (The hand is not shown in the photo for clarity). See the "Tying Tips" section of chapter 3 for more on spinning deer hair.

CHERNOBYL CRAB ■ 65

5. Pull down on the thread steadily, while gradually releasing the hair from the pinch of your fingers. The hair will spin and distribute itself around the hook shank.

6. Press the hair tightly together with a hair packer or your finger tips and thumb. Repeat steps 4, 5, and 6 until the shank is filled. It will probably require two more bunches of hair.

7. Whip-finish and trim the thread. With a razor blade, trim the hair on the outside of the hook (the bottom of the fly) straight toward the bend.

8. Trim the hair roughly on the sides and inside the bend (top of the fly) with scissors.

9. Reattach the thread to the hook behind the eye. Wind the hackle from the rear to the front, palmer style (spiraled). Tie off and trim the hackle.

10. Snip off the hackle fibers from the bottom of the fly.

11. Attach the painted dumbbell eyes just behind the hook eye. Whip-finish and coat the thread.

DEL BROWN'S PERMIT FLY

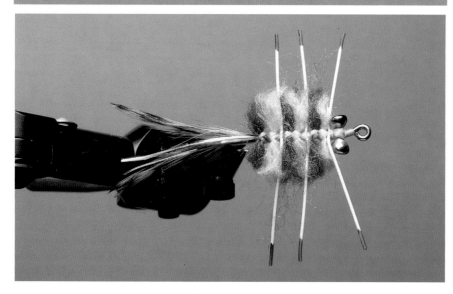

Materials

Hook: standard shank, sizes 2-1/0
Thread: Danville flat waxed chartreuse
Tail: furnace (brown with black center) or Cree (variegated) hackle; pearl Flashabou
Body: tan and brown yarn
Legs: white rubber legs
Eyes: nickel-plated dumbbell eyes

Del Brown was the best known, and probably the best, permit fly fisherman of his generation. He perfected techniques for catching permit on the fly and designed this fly bearing his name. Echoes of Del Brown's famous fly, previously called the Merkin, appear in many crab fly patterns by various tiers. It appears to be more difficult to tie than it is. It is a highly effective fly for any crab-eating species.

Tying procedures

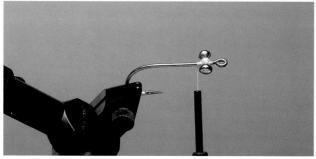

1. Attach thread at the front of the shank. Fasten the dumbbell eyes to the shank securely with crisscross wraps, as for a Clouser or several other flies. Place the eyes close to the hook eye.

3. Tie in several strands of pearl Flashabou between the hackles.

4. Attach a short piece of brown yarn across the hook shank with a couple of X-wraps of thread.

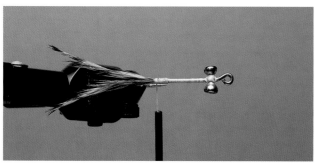

2. Wrap the shank completely with thread. Tie in two furnace- or Cree-colored hackles on each side, curving outwards, at the bend. Before attaching the hackles, as an optional step, you can tie in a small bunch of red fox squirrel hair or bucktail to help separate the hackles.

5. Attach a short piece of tan yarn in similar fashion, close to the brown.

6. Similarly add in a piece of rubber leg material.

7. Repeat steps 4, 5, and 6 two more times until the shank is filled.

8. After tying off and trimming the thread, use the tip of a bodkin to pull the twists from the yarn until it is fuzzy in appearance.

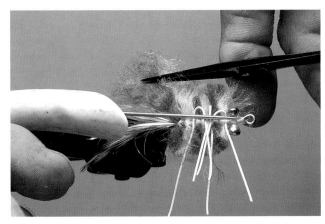

9. Hold the legs out of the way and trim the ends of the yarn on each side. With a red felt marking pen, color the ends of the legs, and coat the head wraps with cement.

Additional tying notes and variations

Olive and dark tan is a common variation.

While tied a bit differently, and using different body material, this Long-legged Crab, tied by Bob Popovics, represents creativity in adapting a concept. It has proven highly successful for striped bass, as well as rays, in the New Jersey surf. It kicks up puffs of sand and shows great leg action as it simulates a crab scurrying through the surf.

Fishing Del Brown's permit fly

The trick is to cast ahead of the fish, let the fly sink, and activate it just as the target fish approaches it. Judging sink rate is critical, and you must adjust the fly's weight and the timing of your cast.

Although known as a permit fly, Del Brown's fly works well on bonefish, as well as striped bass, redfish, snappers, and others.

A fly's sink rate is important when fishing for flats fish (here, a feeding permit). You must present it at the proper depth to intersect the fish's path. You can accomplish this by adjusting the weight in the fly and leading the fish the proper distance when you cast.

ULTRA SHRIMP

Materials

Hook: standard length, sizes 4 through 1/0
Thread: fine clear monofilament recommended
Mouth: Ultra Hair
Legs: saddle hackle
Carapace: Ultra Hair
Eyes: mono burned onto a ball

Shrimps are nearly a universal source of food for inshore ocean fish. Striped bass, redfish, weakfish, seatrout, and other game fish can become selective when keyed to shrimp and crabs, disdaining larger flies. It makes sense to have a few crab and shrimp flies on hand, and few shrimp flies can match this one for realism and effectiveness.

The Ultra Shrimp is another popular design of Bob Popovics, created to imitate the prolific grass shrimp of the east coast. The version shown here reveals significant differences from the original pattern. It is simpler, employs acrylic in lieu of epoxy, and a jig hook, rather than a standard J-hook. Either version can be used as a template and readily adapted to mimic banded shrimp, snapping shrimp, and others. This new design is superior for fishing around grass beds and oyster bars, making it more suitable for bonefishing and redfishing. Also, when at rest it will sit on the bottom, rather than lie on its side, and it imitates a natural shrimp more realistically when retrieved. It also lends itself more easily to weighting, as shown in the variation below.

Tying procedures

1. Using a lighter, melt the ends of two short strands of monofilament, approximately 30-pound test, into a ball, and darken the melted beads with a permanent marker. These will serve as the eyes of the shrimp.

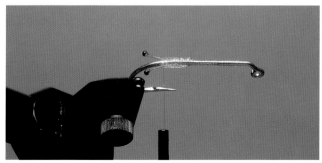

2. Wrap the shank with mono thread, and tie in the two eye stems. The eyes should be about even with the hook bend or protrude slightly beyond.

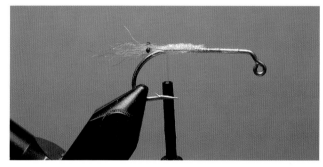

3. Tie in a sparse bunch of Ultra Hair (tan used here) between the eyes and protruding slightly beyond the bend.

4. Tie in a saddle hackle of matching color by the butt.

70 ■ CRABS AND SHRIMP, EELS AND WORMS

5. Wrap several turns of hackle and trim the excess. Pull the hackle back and make several turns of thread around the base of the hackle so the fibers face to the rear. The hackle and Ultra Hair suggest the mouth and feelers of the shrimp.

6. Reattach the hackle again by the butt.

7. Palmer (spiral) the hackle the length of the shank.

8. Rotate the vise or turn the hook upside down. Trim the top and some of the hackle fibers on the sides.

9. Tie in a bunch of Ultra hair across the back of the shrimp to simulate the carapace. Tan is again used here.

10. After tying off the thread with a whip-finish, coat the carapace hair with clear acrylic from the eye to the bend. Expose the acrylic to a blue light, as shown earlier in the instructions for tying the Surf Candy. A second coat is optional. It only remains to trim the excess Ultra Hair.

Additional tying notes and variations

In place of Ultra Hair, you can use Super Hair, Kinky Fiber, Craft Fur, or other synthetics. Note that some are more opaque. The first two mentioned are the most translucent and very realistic in the water.

An acrylic jig hook version with lead eyes attached for weight at the hook angle, incorporating rubber legs for action.

The original and traditional epoxy Ultra Shrimp.

A green epoxy Ultra Shrimp with rubber legs.

Fishing the Ultra Shrimp

Shrimp commonly congregate in vast numbers around grass beds and floating grasses. Cast your shrimp fly to the edges or over the grass, and retrieve with short, staccato strips. Spin fishermen commonly use the Ultra Shrimp as a dropper fly, ahead of a swimming plug. This is intended to simulate a baitfish chasing a smaller prey and to stimulate the competitive nature of gamefish. Some fly anglers use a similar two-fly rig. Drifting the fly beneath a strike indicator on canals or creeks is also popular. Another deadly technique, used for generations, is chumming with natural grass shrimp. Toss a few shrimp into the current every minute or so. When fish follow the trail and come into range, drift back the artificial into the slick.

Weakfish, like the one shown here, striped bass, and many other common inshore species feed heavily on the prolific grass shrimp, especially in bays and inlets. The Ultra Shrimp is one of the most successful imitations you can use.

RABBIT EEL

Materials

Hook:	short or standard length, 2/0-5/0
Thread:	fine, clear monofilament or colored thread to match the body
Tail:	long strip of rabbit fur (Zonker strip); short piece of stiff monofilament
Head/body:	three short rabbit strips; clear 5-minute epoxy
Eyes:	self-sticking prism eyes

Fly catalogs typically list few eel patterns, and fly anglers seldom carry them, and even less seldom use them. Nevertheless, they deserve consideration. The American eel is common in estuaries from the Gulf of St. Lawrence to the Gulf of Mexico. Along the mid-Atlantic and New England coasts, particularly in the later fall, eels are a mainstay of the striped bass's diet. The pattern shown here, really a version of Bob Popovics's Rabbit Surf Candy, is one I evolved, in collaboration with John Zajano. It complements the Keel Eel, a smaller elver imitation shown in the Jiggy section, and it has proven so successful that it is about the only larger eel pattern I use. I generally tie it in chartreuse, olive, or black.

Tying procedures

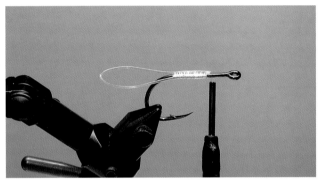

1. Begin by attaching your thread at about the middle of the hook shank. Attach a horizontal loop of stiff monofilament, about 40-pound test. This will reduce the tendency of the tail to wrap around the hook shank.

3. Attach the rabbit strip to the top of the hook shank just behind the eye, and wrap back to the bend.

2. Cut a piece of rabbit strip, 8 to 10 inches long and 3/8 inch wide. Color the hide with a permanent maker to match the color of the fur. A "Lime Peel" Prismacolor pen was used here.

4. Bring the thread forward and tie in a short piece of rabbit strip on top of the hook.

5. Securely wrap the rabbit strip back to the bend.

6. Repeat the procedure with a second short strip of rabbit. Tie it in, starting just behind the place where you tied in the first piece, and wrap to the bend.

7. Repeat with a third, shorter piece, attaching it even slightly farther to the rear. Wrap the whole shank securely; then whip-finish the tying thread and cut it off. Using the same rabbit strip as you used for the tail will assure uniform color, and the small collar of fur behind the head breathes in a fashion reminiscent of an eel's gills.

8. Coat the entire head with 5-minute epoxy. Rotate the fly on a drying wheel or by hand until the epoxy sets.

9. The first coat of epoxy is applied and eyes are in place.

10. This is a close up of the head after marking red gills with a permanent marker and applying a second coat of epoxy.

Fishing the Rabbit Eel

Eel flies should be fished slowly and deep on sinking lines and close to rock structure and weeds, which eels often hug. Striped bass generally feed on eels where there is good current action, such as inlets. Night is definitely the best time for fishing eel flies. I prefer to work the fly with a slow, steady, two-hand retrieve. If this doesn't produce, a slow one-hand retrieve with short, sharp, intermittent tugs can trigger a quick response.

WORM FLY

Materials

Hook: standard shank, size 2-1/0
Thread: black
Tail: red (or orange) marabou "blood" feather
Body: bright red (or orange) chenille
Head: black Ice Chenille or Cactus Chenille or ostrich herl

Most anglers think of worms as freshwater bait only, but major hatches of worms are important for striped bass, weakfish, bluefish, tarpon, and other species. Whether known as sea worms, sand worms, cinder worms, palolo worms, or bearing some other moniker, worms are more prevalent than most anglers realize. For example, in the northeast, spawning hatches of cinder worms occur at night during periods of the new moon and full moon, generally in May, June, and July. They emerge from the mud near marshes and estuaries and drift along in the tide, wiggling, but not swimming like baitfish. Local areas have developed their own favorite worm patterns. Most are about two or three inches long, orange and red being the most common colors for imitations. The pattern shown here is a popular one in New England.

Tying procedures

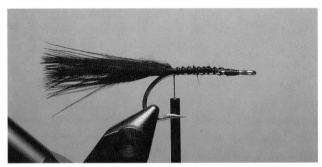

1. Start with the thread attached at the head. Fasten down a bunch of red marabou along the shank so that the ends stick out the length of the hook shank.

3. Wrap the chenille around the shank toward the eye, leaving about 3/16 inch for the head. After tying off the chenille and trimming the excess, attach a strand of black Ice Chenille, Cactus Chenille, or ostrich herl.

2. Tie in a few strands of gold Krystal Flash at the tail over the marabou. Next attach a piece of red chenille at the bend, and then take the thread forward.

4. Make several turns of the chenille, tie down, and trim. Now whip-finish and cement the head.

Variations and adaptations

This is another version of the worm shown above. The hook is inserted inside the hollow chenille and exits from the bottom.

This Palolo Worm Fly, common on tarpon flats, uses orange calf tail in place of marabou, a fluorescent orange chenille body, and a tan chenille head.

Jack Gartside, the fly-tying wizard from Boston, tied this imitation of the ubiquitous sea worm, one of his Wiggle Worm patterns. He also makes this fly, as well as smaller cinder worm flies, with floating and diving heads.

Fishing worm flies

Fish feeding on hatches of these worms can drive fishermen crazy. Often they are unseen, and striped bass in particular may take them in a manner reminiscent of trout taking emerging mayflies near the surface, making small splashes or leaving boils on the surface. The worms wiggle near the surface as they drift with the tide. The best way to fish worm flies is to drift them slowly at, or just below, the surface on a floating line, with occasional twitches.

CHAPTER 6

Topwater Flies

Not all saltwater species feed on top, but nearly every individual who wields a fly rod wishes they would. Fishing sinking lines is mostly a matter of feel, and the underwater take stimulates the imagination. There is a degree of mystery about it. It's a mind game. But the sight of a bluefish, striped bass, jack, snook, or redfish exploding on a surface lure creates a reaction in the angler like nothing else. It's a whole other dimension of excitement, and many anglers are willing to forego the opportunity to catch more, or larger, fish underwater, to experience the visual thrill of taking saltwater game fish on surface flies.

In this chapter I've included six basic top-water designs. As a group, they cover just about any surface fishing situation you might encounter, but they all behave and fish a little bit differently. Each will compel you to hone your fishing techniques. We just never know when fish will prefer the subtle wiggle of the Gurgler or the erratic splashing of a Crease Fly, and only trial and error will tell.

BOB'S BANGER

Materials

Hook: extra long shank hook, sizes 1/0-4/0
Thread: size D rod wrapping thread
Body: Livebody foam cylinder; reflective metallic tape
Tail: bucktail
Collar: Estaz or Cactus Chenille
Eyes: large, self-sticking decal eyes

Bob Popovics employed a number of clever ideas when he conceived this deceptively simple creation. The Banger possesses color, flash, durability, action, and noise, yet its construction requires no shaping, sanding, painting, or gluing. It also stands up to the savage jaws and teeth of bluefish, because the soft head yields to pressure. This cushioning effect protects the tape from sharp teeth, while the tough tape in its turn protects the soft foam from cuts. I've caught numbers of huge bluefish on a single fly. The indentations from powerful jaws and teeth gradually diminish and have little effect on the fly's usefulness or effectiveness. Due to the Banger's modular construction, if the popper head gets damaged, it's a simple matter to replace it, using the same hook. Conversely, if bluefish chew the bucktail, the same head can be inserted on another dressed hook.

Tying procedures

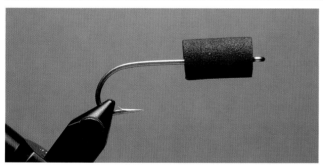

1. After heating your bodkin with a lighter and burning a hole through the center of the foam cylinder (see Popper instructions below), insert the cylinder on the shank to gauge where to start your thread. Attach thread at about the midpoint of the hook shank.

2. Tie in a bunch of medium long bucktail, at least the length of the hook, on top of the shank. Wrap the shank up and back with the heavy thread.

3. Attach a piece of Estaz or Cactus Chenille at the back to add sparkle to the fly.

4. Wrap the sparkle forward to cover the thread. Tie off and trim the material. Then wrap another layer of thread on the shank, and whip-finish. Coat the thread with a heavy layer of head cement.

5. Wrap self-adhesive reflective tape around the foam body cylinder, overlapping at least an extra half wrap.

6. Slip the body over the hook shank. Do not glue in place. Finally, put a spot of cement on the back of the decal eyes, and press them into place.

Additional tying notes and variations

In place of Estaz, you can wrap a collar of hackle or ram's fleece for a different look. Prepare extra heads of different sizes and colors, which you can interchange as needed. If the hole in the foam cylinder wears larger from extensive use, you can always simply coat the shank with glue and fasten the head in place to extend the popper's life. A banger head can also be inserted on the leader ahead of nearly any fly to make an impromptu popper.

This Banger with a larger cylinder head uses fleece in place of the Estaz.

Fishing Bob's Banger

This bug works most effectively when attached to a leader using a nonslip mono loop knot. It never "crabs" or pulls out of line. In fact, when test trolled at fast speed it pulled straight and true. Make repeated, strong, extra long pulls with your line hand. This will pull the Banger beneath the surface. It will leave a wake of bubbles and pop back to the surface. The action is terribly enticing to most predator fish. As an alternative, use moderately fast two-hand strips to keep the bug skipping and hopping across the surface.

Bob Popovics releases a striper caught on his Banger at Lobsterville Beach, Martha's Vineyard, Massachusetts, just before dark.

This bass took a Banger in Pleasant Bay, Chatham, Massachusetts.

CREASE FLY

Materials

Hook: long shank 1/0-3/0
Thread: choice
Tail: Bucktail, Flashabou
Body: sheet foam, colored with permanent marking pens or covered with silver transfer tape; epoxy coating optional
Eyes: self-sticking prismatic eyes
Glues: Loctite (or other cyano-acrylate such as Zap-A-Gap or Krazy Glue)

Captain Joe Blados, a guide and artist from Long Island, came up with this popular but unusual fly. With its blunt nose, it looks unfinished, yet the wider, flat profile gives a good imitation of a "peanut bunker" (small menhaden), small herring, sardine, or pilchard. And what the Crease Fly lacks in aesthetics, it makes up for with an awesome success rate. This fly has earned its stripes, in salt and fresh water. Resembling a wounded baitfish, it flutters, pops, and flops on the surface, a combination popper and streamer.

Tying procedures

1. Attach the thread near the bend of the hook. Tie in a tuft of bucktail, about the length of the hook or slightly shorter.

3. Cut a template as shown from the foam sheet.

2. Add complementary bucktail colors and a few strands of Flashabou. Wrap the thread up and down the shank in crisscross fashion. Whip-finish the thread.

4. Measure the length of the template against the hook shank. It should come just over the hook eye.

5. Run a bead of glue (Loctite CA gel used here) along each side of the shank. Press the foam against the side of the hook shank.

6. Fold the foam over and, starting from the rear, press the sides together along and just below the shank.

7. The finished body is shown here.

8. Add an eye to each side. With marking pens, add a gill, color the front 'mouth' and the back. You can also darken the back and add body markings if you prefer. Coating the entire fly body with a light coating of 5-minute epoxy is also an option.

Californian Ron Dong ties a devastating, yet highly durable, version of the original Crease Fly. It is simply referred to as the RDCF. This fly is more heavily dressed, usually on a larger hook; braided material covers the body, and the mouth is spread open for better popping.

Sheet foam with sparkle and glitter effects is available from tying material suppliers to yield all sorts of finishes. You can also mix glitter flakes into the epoxy.

Additional tying notes and variations

Cut the body templates to suit your whim. You can simulate a variety of baitfish body shapes. When you fold and press the foam, don't press the front completely closed. Try to keep the mouth open for better popping. Also, be careful to keep the hook shank near the bottom of the folded foam body, so that you don't sacrifice the hook gap.

Rick Pope, president of Temple Fork Outfitters, with a 30-pound jack crevalle that he caught on Ron Dong's Crease Fly, Port O'Connor, Texas.

Fishing Blados's Crease Fly

Use one-hand strips for erratic action, to make the lure flutter. Twitch, pause, and pop it, so that it flutters like a struggling baitfish injured or taking its last gasps. I've also had excellent results on false albacore in North Carolina and Florida, fishing the fly underwater on a heavy, fast sinking line. Strip, and the line pulls the fly down. Pause, and the buoyant fly starts to work toward the surface. Instead of attracting fish by surface activity, this action effectively replicates the struggling motion of an injured baitfish trying to reach the surface. Famed author/angler Nick Curcione, a close friend of Blados, used a huge version of the Crease Fly to lure pike in northern Saskatchewan, where it proved deadly. Probably it resembled a wounded sucker or whitefish, both common pike forage.

Surface-feeding striped bass, bluefish, and large jacks, like this one, find the Crease Fly's erratic motion and popping action irresistible.

FLOATING MINNOW

Materials

Hook: standard length, sizes 2-2/0
Thread: personal choice
Wing: red Krystal Flash; several colors of bucktail
Body: foam bug bodies
Eyes: self-sticking prism eyes or small 3-D eyes

Surface lures generally simulate swimming, feeding, fleeing, or wounded and dying baitfish. The Floating Minnow fits this last group. Bob Clouser set out to simulate the posture and behavior of minnows feebly struggling and gasping at the surface. He succeeded beyond expectation. This lure gives an additional option to the more typical popping and chugging surface baits. It sits at the surface at about a 45-degree angle, with its nose breaking the surface, while the weight of the hook sinks the back end. The bucktail imparts subtle action underwater, like a wiggling fish tail.

Tying procedures

1. Attach thread just ahead of the bend of the hook and tie in a small bunch of red Krystal Flash. Tie in a small bunch of bucktail (white used here), about twice the hook length, keeping the hair on top of the hook.

2. Add some gold or silver flash on top of the hair, and tie in another light bunch of slightly darker bucktail, the same length, ahead of the flash, keeping the hair on top of the hook.

3. Add a third color bucktail, the same length, again on top of the shank, finishing all the tying just slightly ahead of the midpoint of the hook shank.

4. Wrap the thread up and back several times along the remaining hook shank. This will give a surface for the cement to adhere to. Whip-finish and cut off the thread. Put a light layer of CA glue over the thread wraps. Loctite gel is used here, but any fast drying cement will work. Gel cements have less tendency to run and are neater to work with than liquids.

FLOATING MINNOW 83

5. Press one bug body against the side of the hook, with the small tapered end forward, and hold in place for 10-15 seconds.

6. Put some CA cement on another bug body and press the two halves together, holding them in place 10-15 seconds until they set.

7. Add either self-sticking prism eyes or small 3-D bead eyes, using a spot of CA cement to hold them in place. Add a light coat of 5-minute epoxy over the entire body (optional). Brush-on CA makes the job easy.

Additional notes and variations

For a different look, you can reverse the bug bodies so that the smaller pointed end faces to the rear.

The Clouser Brightsides Minnow is tied essentially the same way as the Floating Minnow, but with sparser dressing and an elongated body. It appears and floats slightly differently. Captains Brian Horsley and Sarah Gardner, the great husband-wife guiding team from North Carolina, were among the first to use this fly and reported excellent results on false albacore and striped bass.

Fishing the Floating Minnow

Cast, let it sit, and twitch lightly periodically. I fished with Bob Clouser when he put on a show, catching false albacore at will near Cape Lookout, North Carolina. He threw it into pods of albies that were crashing through schools of bay anchovies, and simply let it sit and bob, with no additional action. I can only assume the lure's posture in the water had something to do with its success. Regardless of the cause, the albacore gobbled them like popcorn.

When albacore or stripers are boiling under baitfish like this, it's wise to simply cast and let the fly float in the area with no additional action. The fish appear to pick up crippled baits that they can catch with little effort.

GURGLER

Materials

Hook: regular or long shank hook, sizes 1-2/0
Thread: flat waxed
Body and carapace: Livebody or other closed cell sheet foam
Tail: bucktail; sparse flash
Rib: palmered saddle hackle

From the vise of Bostonian tier Jack Gartside comes the Gurgler; not a true popper, not a slider. He claims, "It works well on shy fish which might otherwise be spooked by a large, noisy popper." Amen. I first tried it under precisely those conditions. Fishing with Captain Dan Marini on Pleasant Bay in Chatham, Massachusetts, I watched bass rise toward my poppers then veer off when I activated them. Changing color and size failed to change results. I switched to a Gurgler, not hopefully I might add. I cast it, let it sit for what seemed too long, and then gave the line slight twitches. The fly shimmered and wiggled, generating subtle waves as it inched along. That was clearly the right course of action. Stripers crashed the bug time and again. It has become a fly box staple. And, it's easy to tie.

Tying procedures

1. Attach thread at bend of hook. Tie in a long, sparse bunch of bucktail, and add several strands of Flashabou to complement the color hair you select.

3. Cut off the tip and trim some of the barbules from a saddle hackle, and attach it to the hook, as shown, with the dull side to the rear.

2. Tie in a single strand of wide pearlescent Flashabou and leave it trailing to the rear for now.

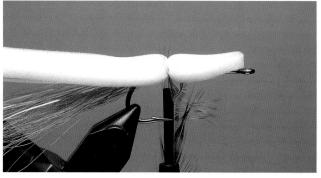

4. Cut a strip of closed cell foam, about ⅝ inch wide and several inches long. Tie the foam down tightly to the hook shank at the bend.

GURGLER ▪ 85

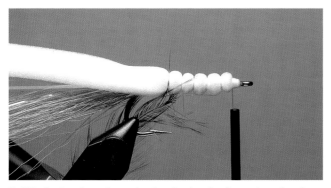

5. Work the thread up the hook shank, fastening the foam down tightly into five segments.

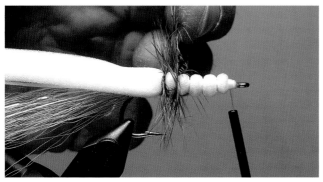

6. Now wrap the hackle forward in between the body segments.

7. Tie the hackle off ahead of the body and trim the excess.

8. Pull the foam strip forward from the bend, and tie it down behind the hook eye.

9. Pull the wide pearl Flashabou strip forward and tie it down in the same place. Trim the Mylar and the foam, leaving about $1/4 - 3/8$ inch protruding forward.

Additional tying notes and variations

Black, yellow, white, chartreuse, and combinations of other colors all have their fans. Experiment.

Fishing the Gurgler

This fly was designed for subtle action. Work it slowly, with short, intermittent strips. Pull it a little harder, with a continuous, easy, long pull and you generate a noise that gave the lure its name.

POPPER

Materials

Hook: Long shank, 1/0-3/0
Thread: flat waxed or personal choice
Body: preformed foam
Tail: bucktail, Flashabou, or other tinsel
Eyes: self-sticking eyes, fairly large

This is a generic surface "popping bug," and the first I would choose if limited to only one surface fly-rod lure. Different versions are known by many names. In New Jersey it was the venerable Ka-Boom-Boom striped bass popper. The KBB had a deeply cupped face, great buoyancy, and produced a commotion out of proportion to its size. Unfortunately, it required a lot of work, cutting or drilling, shaping, sanding, gluing, and painting the cork body. Although I have a strong nostalgic attraction to the KBB, I rarely bother with making them, opting instead for the quick and easy foam body popper shown here. Any novice tier can make this fish-catching surface lure with little talent or practice. I believe that the bug works better if I ignore the predrilled hole, and burn a hole at the bottom of the foam body, as shown below.

Tying procedures

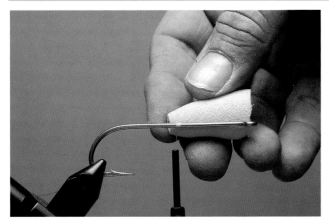

1. Measure the body against the hook. It should be approximately one-half the shank length to avoid narrowing the gap and impeding hooking. Starting with the thread attached at a point just ahead of the end of the body will allow you to make a nicely finished bug, with the wraps covered by the body.

2. Tie in a medium-size bunch of bucktail. Add some Krystal Flash, Flashabou, or other tinsel for a touch of flash. Cross-wrap the hook shank with thread and whip-finish. This provides a base to hold the cement.

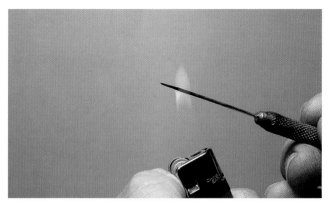

3. Heat the end of a bodkin with a lighter.

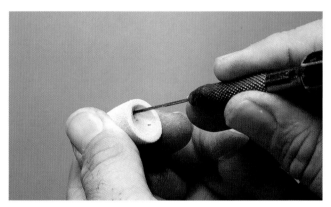

4. Insert the bodkin into the face of the body near the bottom, and push it through so that it exits out the rear hole. Work the hot needle around as you push it, to enlarge the hole slightly.

5. Coat the thread wraps with a fast-drying CA cement or 1-minute epoxy. This will insure that the body won't turn on the hook.

6. Push the body onto the shank, being careful to make sure the hook eye exits from the hole you burned, not from the original hole in the center of the body. Attach a self-sticking decal eye on each side of the body. A spot of epoxy on the back will help secure the eyes in place.

Additional tying notes and variations

This assortment of poppers covers a wide spectrum of cork-, foam-, and balsa-bodied lures. Some involve more work than the simple one shown above. Along with the other designs in this chapter, they will give some idea of the imagination and creativity of tiers and anglers.

A cork body Ka-Boom-Boom. Its construction calls for cutting, drilling or slotting, sanding, gluing, and painting a cork form. Over the years I've caught more fish on this popper than any other, from striped bass in New Jersey to peacock bass in South America.

The little Dink Popper has sparse dressing. It is a popular and effective popper for redfish when they want noise but are intimidated by larger lures.

The Potomac Popper was a smallmouth bass bug that Lefty Kreh popularized. It has a flat bottom and flat, slanted face. John Zajano created this saltwater version. Using a sanding wheel, he shaped the body from a dense foam cylinder of Livebody. A slot is cut in the bottom and the hook glued into place. It's light, pops great, and is easy to cast.

This large Spouter-type popper is a commercial lure. It has a hole running from the face to the top of the body through which water is forced on the retrieve, making a spray and additional surface commotion.

Fishing Poppers

Tie a popper to your leader with a loop knot. Let the popper sit for a second or two after casting, and pull out the slack in your line before moving the lure. With your rod tip low, pointing toward the bug, give the line a sharp tug, stripping it about a foot, and causing the bug to make a splash or boil on the surface. Pause, and repeat the action. You can experiment with louder or softer pops and shorter or longer strips, but this is the basic approach.

I don't know whether bluefish love or hate poppers, but they attack them with abandon when chasing schooled baitfish. These are the first I caught on Martha's Vineyard over a quarter of a century ago.

Angler/writer Nick Curcione and Captain Dan Marini display a striper that fell for a large black popping bug in Chatham, Massachusetts.

SLIDER

Materials

Hook: long shank, preferably with hump shank, sizes 2-3/0
Thread: your choice
Tail: Kinky Fiber, saddle hackles
Flash: Flashabou or other tinsel
Collar: saddle hackle or wider schlappen hackles
Head: tapered foam premade form
Eyes: self-sticking prism eyes

Like the Bendback Fly, the Slider has no prescribed dressing, save a tapered, bulletlike front. You can fashion a slider simply by reversing the head used for the Popper (see "Popper" instructions), or be as creative as you like. The version shown here is a pattern of John Zajano. The Slider slips through water with movement and subtle disturbance, but little real commotion and noise. There are times when fish simply prefer the quiet approach.

Tying procedures

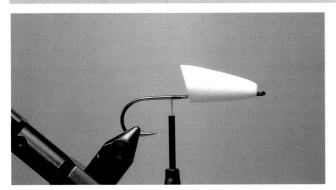

1. Cut a slot in the bottom of the head with a razor blade. To gauge where to attach the thread, insert the head over the hook eye. Attach the thread to the hook shank at about the midpoint of the shank.

2. Remove the head. Tie in a bunch of Kinky Fiber about twice the length of the hook shank. Over this, tie several strands of Flashabou tinsel the same length as the tail. You can mix the colors.

3. Tie in two long hackles on either side of the tail, dull sides in.

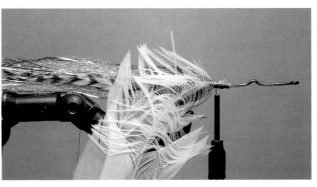

4. Tie in two hackles by the tips, dull sides in. Schlappen hackles are shown here to give more bulk, but saddle hackles can be used as well.

5. Wrap the hackles several turns around the hook. You can tie them one at a time or both together.

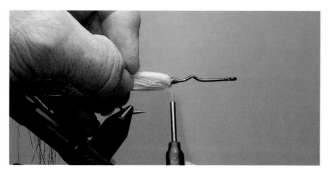

6. After tying down the hackle and trimming the ends, pull back the hackle with your fingers and wrap the thread around the base to hold the hackles curving to the rear.

7. Put some cement (Loctite or other) in the slot in the head; then fit the head over the hook shank and run another bead in the slot. Squeeze the head for a few seconds until the cement sets.

8. The completed Slider. It simply remains to add the eyes and any other markings you want with marking pens. Optionally, you can brush the head with epoxy, as for the Crease Fly and the Floating Minnow.

Additional tying notes and variations

This commercial slider shows how you can make a slider type by simply reversing the head used for a popper. (See "Popper" section.)

The Slim Jim slide/popper is the essence of simplicity: a tail of deer tail, a few wraps of Cactus Chenille, and a foam cylinder, decorated with prism eyes. It's a great small fly for picky striped bass. In red it makes an acceptable worm fly. (See "Worm Fly" section in the previous chapter.)

Fishing the Slider

My two favorite times for fishing sliders are after dark, when their subtle waking motion attracts fish prowling shallow waters, and when large striped bass look for herring or menhaden in the spring. The latter calls for flies of perhaps 8–10 inches.

You can experience exciting surface action for striped bass when fishing Sliders in a flat surf, particularly during periods of low light after daybreak and at dusk.

CHAPTER 7

Bonefish Flies

History, or tradition, has it that bonefishing, as we know it, i.e., casting to feeding and tailing fish in shallow water, started shortly before World War II. The sport picked up momentum in the 1960s, and really took off in the 70s and 80s. Today, I suspect that saltwater anglers probably make more destination trips for bonefish than any other species, to the Keys, Bahamas, Mexico, Caribbean, Christmas Island, Seychelles, and other exotic venues. I suppose it's also a fair analogy that bonefishing is to saltwater what trout fishing is to freshwater angling. Nevertheless, because bonefishing lacks the diversity of trout fishing, there really isn't much variety in fly design, and the phrase "bonefish fly" conjures up a stereotypical design in the minds of most anglers. As discussed earlier, it is difficult to classify saltwater flies according to the species they are designed to catch. Bonefish and, only to a slightly lesser degree, tarpon flies are the prime exceptions. Not that other species can't be and aren't caught on "bonefish" flies, but the flies described here were all created with this one fish in mind.

I've included in this section some of the standard patterns on which anglers have relied for the last couple of decades. They have accounted for thousands of bonefish, and nearly all are quite simple to tie, even for beginners. Tom Gilmore, author of several angling books, advises anglers to take a small assortment of bonefish flies on trips, just for starters, along with a compact tying kit. With limited materials and experience, you can tie any flies you need in a few minutes, on site.

From the seemingly endless array of bonefish flies in shops and catalogs, I've selected a handful that will provide sufficient options for the fussiest fish, or tier. Collectively they offer sufficient options to catch bonefish anywhere in the world. You needn't carry all of them. From the small assortment included, you should be able to find several that will strike your fancy and that you enjoy tying. For example, if limited to two, you might select a Gotcha and a Swimming Shrimp. The first is light in color and sinks readily; the latter is darker, splats on water, and has a deer hair body which sinks and moves differently. Similarly, a group including a Mini Puff, a Crazy Charlie, and a Snapping Shrimp represents a practical assortment that would cover a spectrum of needs and situations. Take such an approach when selecting flies to tie for any other species as well.

For various reasons, I have placed several important patterns that might be classed as bonefish flies in other chapters. Tim Borski's Chernobyl Crab, Del Brown's Permit Crab, and Bob Popovics's Ultra Shrimp are found in the chapter devoted to crab, shrimp, eel, and worm patterns. I consider these versatile and more generally useful designs, suitable for permit, redfish, striped bass, seatrout, and many other fish, in addition to bonefish. Also, the small bonefish Clouser, a simplified Clouser Deep Minnow, needs no special instructions, and I show it under that heading.

Bonefish eat almost any living thing in their domain, provided it's relatively small, like shrimp, crabs, various minnows, snails, worms, and clams. Hooks for bonefish flies range from size 8 up to 1/0, size 4 being perhaps the most common. Nearly all bonefish flies are now tied "upside down," that is, with the hook riding up. Early bonefish flies, like the Phillips Pink Shrimp, tended to hang up in shallow water or catch on coral growths or grass. Many of these, such as the popular Frankee Belle, have now been redesigned, using the same color schemes, to ride hook up. Another important consideration for bonefish flies is sink rate. Fish feeding on the bottom or in the weeds will miss seeing a fly that rides too high in the water column, and flies that sink too readily may hang up excessively. Vary your flies' sink rates by varying the size of the lead or bead chain eyes and the buoyancy of the materials.

BONEFISH SPECIAL

Materials

Hook: standard length, sizes 2-6
Thread: black
Tail: orange marabou
Body: clear monofilament over flat gold tinsel
Wing: white calf tail or bucktail, grizzly hackle tips

Floridian Chico Fernández created the Bonefish Special, and Lefty Kreh says, "It's one fly every bonefisherman should carry." I believe the contrast of the black/white wing, the tinsel, and splash of bright orange make this fly appealing.

Tying procedures

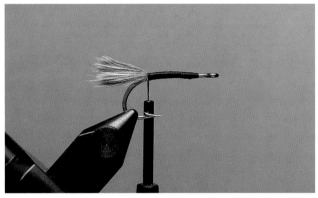

1. Attach thread close to the eye, and tie in a short tuft of orange marabou. Tie down the stem of the marabou along the whole length of the hook shank. This will make a smooth underbody.

2. Attach a piece of clear mono, tying it down along the length of the hook shank. Take the thread forward and tie in a piece of flat gold tinsel near the eye of the hook.

3. Wrap the tinsel down the shank and back to the front.

4. Tie off the tinsel and trim the excess. Wrap the mono forward over it, and tie that off.

BONEFISH SPECIAL ■ 93

5. Tie in a white calf tail wing, reaching a short distance past the bend of the hook.

6. Add a grizzly hackle point on each side of the wing. Finish shaping the head with the thread, whip-finish, and coat with head cement.

Additional tying notes and variations

Make sure that the under layers of body material—the thread, the end of the mono, the tinsel—are level and smooth, with no bumps. Otherwise you will not be able to neatly wrap the mono over the tinsel. The hackle point feathers may be tied curving in or out.

This early bonefish fly, first tied by guide Lee Cuddy, makes effective use of the popular red/white color combination.

The Frankee Belle was named after two of the first Florida Keys women guides. This very old bonefish pattern is now commonly tied reverse, i.e. with the hook riding up.

Similar to the Bonefish Special, this Bimini Bone fly sports a pink body.

CRAZY CHARLIE

Materials

Hook: standard length, sizes 2-6
Thread: white 3/0 flat thread or Monocord
Tail: silver Flashabou
Body: clear mono, Stretch Tubing, or V-Rib
Eyes: silver bead chain
Wing: white hackle points

Bob Nauheim tied the first "Nasty Charlie" (later renamed "Crazy Charlie") to imitate small glass minnows on Bahamas flats. The original pattern, which I show here, featured hackle point wings splayed outward. Nowadays, a calf tail wing version appears to be far more popular, and tiers also substitute pearl or silver flash wings. The fly is also tied in a variety of colors. It's safe to say that the Crazy Charlie, with its family of derivatives and variations, probably represents the most widely used bonefish fly in the world today.

Tying procedures

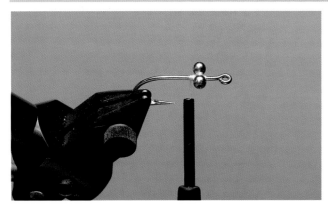

1. Starting with the thread attached about a third of the way back on the hook shank, secure bead chain eyes to the shank with crisscrossed thread wraps.

3. Attach a piece of clear V-rib or monofilament at the bend; then take the thread forward and tie in a piece of flat silver tinsel behind the eyes.

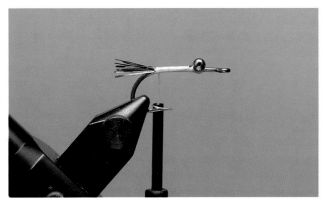

2. Take the thread to the bend and tie in a short tail of silver Flashabou sections.

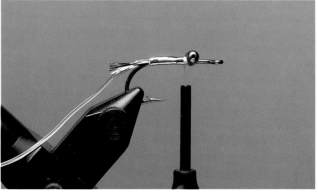

4. Wrap the tinsel to the rear, then back to the eye; tie it off, and trim.

CRAZY CHARLIE

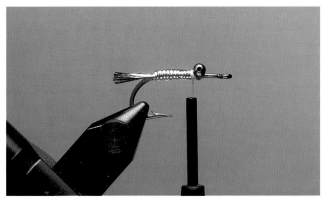

5. Wrap the V-rib or mono forward over the tinsel; tie it off behind the eyes, and trim the excess.

Pink, as well as olive and tan, are popular Crazy Charlie colors.

6. Add four white hackle point tips, dull sides out, as a wing. Finish the head with additional thread wraps and whip-finish.

Although the flash wing Crazy Charlie is normally tied with Flashbou, I tied this one with Krystal Flash for a different effect.

Additional tying notes and variations

The Crazy Charlie with a calf tail wing and pearl flash is now more popular than the feather wing version. Olive, pink, and tan are other favorite colors.

GOTCHA

Materials

Hook: standard length, sizes 2-6
Thread: 3/0 pink flat nylon floss
Eyes: silver bead chain, size dependent on weight and sink rate desired
Tail: pearl braided Mylar tubing
Body: pearl braid
Wing: tan Craft Fur; Krystal Flash

The Gotcha's popularity has risen like a meteor in recent years. Perhaps the pink coloration is attractive to bonefish—or fishermen. Jim McVay created it as a shrimp imitation, especially for Bahamas fishing, but it is a top choice just about everywhere anglers cast to bonefish. It seems that everyone, myself included, makes minor variations to the pattern, but the widespread success of most of them attests to the efficacy of the basic design.

Tying procedures

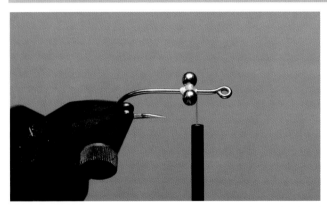

1. Attach thread about a third of the way back on the hook shank. Secure bead chain eyes to shank with crisscrossed thread wraps, as shown above for the Crazy Charlie.

3. Unravel the braid with the point of a bodkin.

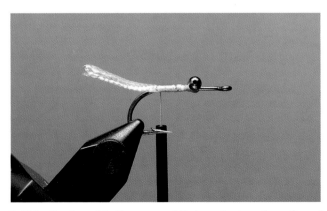

2. Take the thread to the rear and tie in a short section of pearl hollow braid.

4. Tie in a longer section of the same braided material at the bend.

GOTCHA ■ 97

5. Wrap the braid to just behind the eye and tie it off. Move the thread ahead of the eyes.

6. Tie in up to a dozen short strands of pearl Krystal Flash.

7. Tie in the Craft Fur wing, shape the head with tying thread, and whip-finish.

Additional tying notes and variations

Most tiers commonly make the pink head on the Gotcha a bit longer and larger than normal. Orange craft fur is sometimes substituted for the tan wing when fishing over dark bottoms.

This Gotcha is tied without body and tail. I have had excellent luck with this simplified version.

This darker version, with an orange wing, is most effective over darker bottoms.

HORROR

Materials

Hook: standard length, sizes 4-8
Thread: 6/0 black or red
Wing: natural brown bucktail
Body/head: fine yellow chenille

The Horror is one of the first, if not the first, bonefish flies tied in reverse with the hook bend up, which has since become the standard. It achieved this without weighted eyes by relying on the buoyancy of its bucktail wing, tied inside the bend. Dick Brown tells the story of how and why Pete Perinchief of Bermuda came up with the design. The dressing is so simple that tiers feel compelled to dress it up with feathers or a full chenille body. Nevertheless, the original pattern, as shown here, still works just fine. I'm partial to the Horror because I caught my first bonefish on it in Belize more than thirty years ago.

Tying procedures

1. Attach thread to the hook close behind the eye. Tie in a short wing of bucktail about one-third of the way back from the eye.

3. Wrap several turns of the chenille ahead of the wing and tie it off.

2. Scrape the fluff from the end of a short piece of fine yellow chenille, and attach it behind the eye.

Additional tying notes and variations

One wrap behind the wing can help to prop and support it.

Some anglers feel that the Horror looks incomplete, so they opt for a full yellow chenille body.

MINI PUFF

Materials

Hook: standard length, sizes 4-8
Thread: pink
Wing: tan calf tail, grizzly hackle points
Body/head: pink chenille
Eyes: small silver bead chain

The Mini Puff, like the Horror, is a simple pattern, easy to tie, and calls for a minimum of materials. It is a simplified and smaller version of Nat Ragland's famous Puff Fly that was designed for permit. It appears the original was tied with tan chenille, but pink has become very popular. The chenille around the bead eyes cushions the fly's landing, allowing a softer presentation. I've had excellent results in the Bahamas with this fly, especially in very skinny water.

Tying procedures

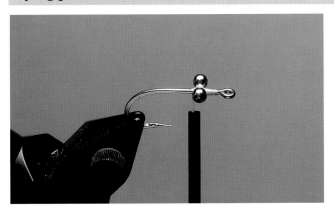

1. Attach thread about a third of the way back on the hook shank. Attach bead chain eyes securely with crisscrossed thread wraps.

3. Tie a grizzly hackle point on each side of the calf tail.

2. Tie in a small clump of tan calf tail as a wing, in front of the eyes.

4. With your thumbnail, strip the fluff from the end of a piece of fine pink chenille and attach it by the core thread in front of the wing.

5. Wrap the chenille, crisscrossing several turns, around the eyes. Tie off the chenille and whip-finish the head.

This dark brown Puff variation sometimes does the trick when fish shy away or spook from lighter colors.

Additional tying notes and variations

The grizzly hackle points in the wings are sometimes attached curving outward, sometimes inward. For fishing extremely shallow water, I always tie up several flies without the bead eyes. Like The Horror, these sink very slowly and prevent hanging up on coral or turtle grass.

A Mini Puff is tied here with tan chenille. White is also used.

SNAPPING SHRIMP

Materials

Hook: standard length, sizes 2-6
Thread: black 3/0
Body: light tan yarn, with orange butt
Wing: brown FisHair

Chico Fernández created this imitation specifically to simulate the snapping shrimp. It's a basic, simple-to-tie bonefish fly. It's a classic pattern and one of the first to try when you don't need a weighted fly.

Tying procedures

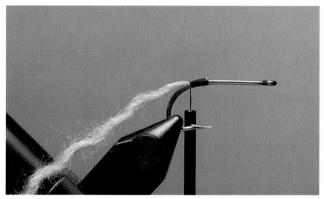

1. Attach thread at the bend of the hook, and tie in a piece of orange yarn (Sparkle Yarn used here).

3. Attach a piece of tan yarn, tying the butt end along the hook shank for extra thickness.

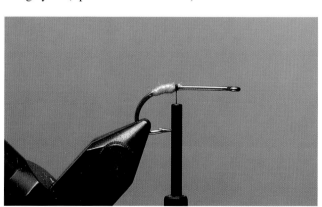

2. Wrap several turns of the yarn and tie off.

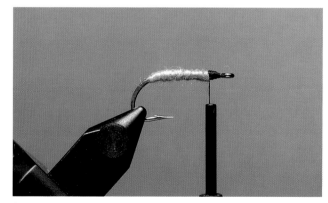

4. Wrap the yarn forward, and tie off close to the hook eye.

5. Tie in a wing of FisHair or Craft Fur, wrap a tapered head, and whip-finish.

The yellow body and mottled effect of Cree (brown, black, white) hackle points on this fly make a nice option to the plain Snapping Shrimp.

Additional tying notes and variations

A small eye may be painted on the head or a small prism eye added. Dubbing may be substituted for the yarn of the body.

The Snapping Shrimp often is tied with a lighter-colored wing and dark barring added with a felt marker.

SWIMMING SHRIMP

Materials

Hook:	standard length, size 2-6
Thread:	white flat thread
Tail:	tan calf tail; gold Krystal Flash
Wing:	grizzly hackle points
Body:	tan deer body hair
Eyes:	lead eyes, white iris and black pupil
Weed guard:	stiff 20 lb. monofilament (optional)

This Tim Borski creation is a bit more involved to tie than any of the previous patterns, but it's worth the trouble. This fly has more subdued coloration than most bonefish flies. More importantly, it enters the water differently. The deer hair body lands with a subtle splat, as opposed to the plunk of more skimpily-dressed flies. I consider this fly essential; it will add variety to your bonefish arsenal. Review "Tying Tips" in chapter 3 on spinning deer hair.

Tying procedures

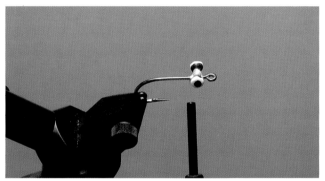

1. Attach thread near the hook eye. Fasten painted lead eyes to the shank with cross wraps. Apply a drop or two of head lacquer to secure.

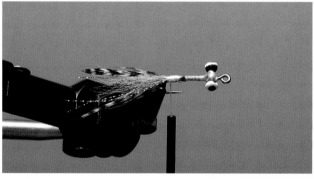

3. Add a few strands of gold Krystal Flash on each side of the tail, then tie in one grizzly hackle tip (curving in or out) on each side.

2. Tie in a small bunch of tan calf tail, wrapping the butts along the shank to create a smooth underbody.

4. Clip a small bunch of deer body hair from the hide, even the tips with a stacker (see "Tying Tips," chapter 3), and clip the butt ends. Fasten the hair loosely to the hook with two wraps of thread.

5. Now, while lightly holding the hair between your thumb and forefinger, pull down on the thread. The hair will "spin," distributing itself around the hook, and flare out as you pull tighter on the thread.

6. Pack the hair tightly; then work the thread forward and make one or two turns around the shank in front of the hair.

7. Repeat steps 4 through 6 with a second bunch of deer hair. A third bunch may be required, depending on how much you can handle at one time and on the length of the hook shank.

8. The hook shank here is packed with deer hair.

9. Trim the hair close to the shank into a bullet shape with an oval cross section.

Additional tying notes and variations

This version of the Super Swimming Shrimp has all the deer hair on the top of the hook shank and rubber legs for added action.

While not really a variety of the same fly, this variation of Borski's Bonefish Slider is another good imitation for bonefish and other flats fishes that reveals the same kind of thinking: It has a longer tail of synthetic hair, barred with a permanent marker.

Fishing bonefish flies

Captain Lenny Moffo displays a nice bonefish the author caught while being guided in the Florida Keys.

The key to successful bonefishing involves gauging the fly's sink rate, water depth, and speed of the fish.

A small, but well-chosen assortment of bonefish flies can provide enough ammunition for a week of fishing.

Carry an assortment of flies for different conditions. In very skinny water such as shown here, lighter flies that sink slowly and that allow a softer entry will take more bonefish.

CHAPTER 8

Tarpon Flies

Tarpon are widely distributed in tropical and subtropical waters. Flats, lagoons, rivers, and canals in the Florida Keys, Central America, and the Caribbean are the most popular destinations for American anglers. Like bonefishing, tarpon fishing developed on a fast track in the mid-twentieth century and produced a large variety of specialized flies. Only a select few are shown here, but these are foundation flies for most of the important designs used then and now. More recent designs like the Bunny Fly and Toad Fly are widely used, but I've included a few of the old classic "Keys-style" flies for several reasons. One is for a purely historic reason; namely, they have a long tradition. Perhaps more importantly, and certainly more practically, they are very simple to tie, some requiring only one or two materials. You will learn to handle basic materials and how to gauge proportions. If you can tie these, with little additional effort, you will be able to tie nearly any tarpon fly.

Interestingly, many classic tarpon flies lack flash, although some tiers optionally add a bit of tinsel. This on occasion was due to a fear that in clear tropical waters the flash can spook fish. Certainly tarpon flies tied for fishing murky waters regularly employ flash (see the instructions for the Deep Water Whistler, page 115, widely used for tarpon in Central American jungle rivers). A more cogent explanation is that when the earlier tarpon flies were developed, only metallic tinsels were available, and these didn't stand up well to the saltwater environment.

APTE TARPON FLY

Materials

Hook: standard length, 2/0-4/0
Thread: red-orange
Wing: red-orange and yellow neck hackles, wide and curved
Collar: red and yellow saddle hackles
Head: red-orange thread

Legendary Florida angler Stu Apte designed this colorful and popular fly. Like Lefty's Deceiver shown earlier, it was selected by the U.S. Postal Service to grace a 29-cent stamp several years ago. The Apte Tarpon Fly stands as the prototype for nearly all the "Keys-style" flies, with splayed feather wings tied at the bend, leaving a bare hook shank, which is usually wrapped only with thread. This allows freedom of action to the wings and makes the fly virtually foulproof.

Tying procedures

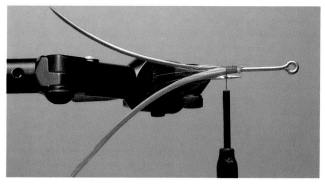

1. Attach thread at the bend, above the barb. Prepare eight hackles: two yellow outside of two orange for each wing. Moisten them and fasten them to the hook shank, curving with dull sides out.

3. Wrap the hackles around the shank so that the flues of the feathers stand perpendicular to the shank. Tie off the feathers with several wraps of thread, and trim the excess.

2. Tie in one yellow and one orange hackle, dull sides to the rear, at the bend of the hook where you attached the wings.

4. Tie in two more hackles and wrap them as you did the first pair.

5. Tie off the hackles and wrap the shank completely with thread. Whip-finish and coat the thread thoroughly with head lacquer.

Additional tying notes and variations

The grizzly and yellow Chinese Claw is another standard Keys-style tarpon fly, tied in similar fashion to the Apte Tarpon Fly.

The Black Death tarpon fly. It is tied in several variations, often with marabou substituted for the hackle. See chapter 1 for other versions of this fly.

The red and white Homer Rhode streamer fly. It was one of the first flies tied specifically for tarpon, and one of the most popular tarpon flies of all time.

This is a commercial version of the Tarpon Glo, designed by Jonathan Olch. It sports highlights of fluorescent orange and green.

BABY TARPON FLIES

These are simply smaller versions and variations of Keys and other standard tarpon flies. In addition to being deadly for "baby tarpon" (approx. 5-30 lbs.), they are highly productive for snook, redfish, seatrout, and other smaller species. Since they require no different tying procedures, I've simply included photos, without specific tying instructions. There are no prescribed dressings for Baby Tarpon Flies. Hook sizes are typically 4 through 1/0, and overall lengths about 2 to 3 inches. Weedguards are not generally advised for tarpon flies, but are useful when fishing for snook and redfish in cover.

This fly uses the color theme of the Apte Tarpon Fly.

This is the Cockroach in its "blue period."

Red and white is ever popular. I've done especially well on snook with a similar pattern.

This is a brown translation of the same fly.

This is a miniature of the Cockroach described on page 112.

This is a particular favorite of Lefty Kreh. It features a wing of dark or light brown marabou and orange grizzly hackles, with gold braided tinsel wrapped along the hook shank.

BUNNY FLY

Materials

Hook: standard shank 3/0-4/0
Thread: black
Body: medium chenille
Wing: strip of rabbit hair
Rib: clear monofilament
Hackle: neck hackles

In recent years, rabbit fur has become popular for streamer flies for salt and fresh water use. It is a durable and lively material that waves and breathes in the water with the slightest movement, giving lifelike action to flies. Many versions of Tarpon Bunny flies are found in the boxes of tarpon anglers. Famous anglers—Mike Wolverton, A.J. Hand, and many others—have their names associated with great rabbit fur creations.

Tying procedures

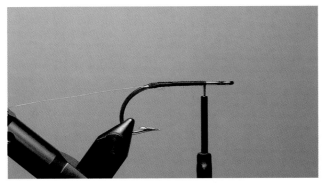

1. Attach the thread at the bend, above the barb. Tie in a piece of 12- to 15-pound mono to the shank and leave it hanging to the rear.

3. Wrap the chenille down the shank and back, and tie it off at the front of the fly. After pulling a little fur from the end of a rabbit Zonker strip to expose the hide, attach the end of the hide at the front, near the eye.

2. After scraping fluff from $1/4$ inch of chenille with your thumbnail, tie the exposed core of the thread at the front of the shank.

4. With your bodkin, part the fur and tie down the skin to the hook with the monofilament. Spiral the monofilament up the body every $1/8$ inch or so and tie off at the head, being careful not to tie down the fur. Wetting the fur lightly while tying helps keep it under control.

BUNNY FLY ■ 111

5. Attach two or three hackles by the tips at the head. You can make the fly neater by tying them in at slightly different spots around the hook shank, rather than all in one bunch. Pheasant feathers were used here, but saddle hackles with long flues or schlappen hackles work well also.

6. Wrap the hackles and cinch them with the tying thread one at a time. Pull the feathers back with your fingers and wrap the thread up against the base of the hackles to fold them back over the body, so that the hackle collar flows into the rabbit wing. Whip-finish and lacquer the head.

Additional tying notes and variations

Instead of using monofilament to rib the body and tie down the rabbit, you can use the tying thread. Attach the chenille at the rear, wrap it forward and back. Tie down the rabbit first at the rear and rib forward with the thread. You can also wrap a few turns of rabbit to form the collar, in place of the hackle.

A highly effective but simple tie, this tarpon fly consists of a dyed rabbit Zonker-cut strip wing, plus a collar of contrasting cross-cut rabbit.

COCKROACH

Materials

Hook: Standard shank 2/0-4/0
Thread: black
Wing: grizzly neck hackles
Collar: gray squirrel tail

Sometimes fish prefer more subdued colors. When a darker-colored tarpon fly is required, the Cockroach is one of the first that anglers turn to. Here the fly is tied Deceiver-style, with a squirrel tail color close to the eye. It is also commonly tied Keys-style, with all materials tied at the bend and the hook shank left bare or wrapped with black thread.

Tying procedures

1. Attach the thread at the bend of the hook, above the barb. Depending on fly size, match 6 or 8 grizzly hackles (3 or 4 for each wing), flared with curved sides out.

3. Take the thread forward toward the eye. Tie in a bunch of gray squirrel tail hair to one side of the hook.

2. Moisten the hackles for better control while tying, and attach them to the top of the hook shank at the bend, as for the Apte Tarpon Fly.

4. Tie in a second bunch of squirrel tail hair on the opposite side. Spread the hairs around so that they are distributed evenly around the fly. Whip-finish and lacquer the head.

TOAD FLY

Materials

Hook: TMC 600 SP, sizes 2/0-4/0
Thread: personal choice, to complement fly color
Tail: marabou
Collar: rabbit strip
Body: twisted yarn
Eyes: plastic or metal bead eyes

The Toad Tarpon Fly, evolved from an earlier bonefish fly, made its entrance on the tarpon scene fairly recently but quickly gained fame for its success and has become a standard among Florida tarpon guides and anglers. Andy Mill, widely regarded as one of the best tarpon anglers ever, gives this fly his highest praise. The earliest version featured a rabbit strip tail, but marabou has become the preferred material of many tiers. Like most tarpon flies, it is easy to tie and readily adaptable to a variety of color combinations. The black and purple fly shown in the steps below is favored when the water is murky or otherwise off color.

Tying Procedures

1. With the thread attached at the bend, just above the barb of the hook, tie in a full marabou tail, about two and a half times the hook length. You may need two feathers to get the necessary fullness. Trim off the butts.

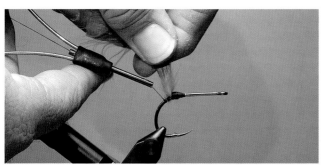

2. Lift the marabou and make several wraps under the tail. This props it for better action and helps to keep it from wrapping around the hook.

3. Tie in a strip of rabbit fur on the hide, and make several wraps around the hook. I prefer cross-cut rabbit when wrapping around the shank, as opposed to a lengthwise Zonker strip, as used on the Tarpon Bunny.

4. Tie off and trim the rabbit. Wetting the fur helps to keep it out of the way when you are tying the body. Tie in two pieces of yarn about an inch and a half long, with X-wraps. Note: This is the same procedure used for the Del Brown Permit Crab.

5. Repeat step 4, adding yarn until you fill the hook, stopping just short of the hook eye. Leave a little space to add the eyes.

6. Whip-finish and cut the thread. With the tip of your bodkin, pull the yarn twists apart.

7. Now trim the yarn on the sides until you have a neat, flat, fuzzy body. This side view shows how flat the head is.

8. Reattach the thread. Add black plastic or metallic bead chain eyes, crisscrossing the thread tightly. Whip-finish, trim, and lacquer the thread wraps.

Additional tying notes and variations

This bright chartreuse and yellow Tarpon Toad pattern is used in exceptionally clear water.

The tan and orange version is a popular middle of the road alternative to the light and dark versions.

WHISTLER FOR TARPON/DEEP WATER

Materials

Hook: standard or shorter shank 4/0-5/0, heavy wire
Thread: usually black
Body: chenille
Wing: wide saddle or neck hackles
Collar: bucktail
Flash: Mylar tinsel
Eyes: large bead chain
Underbody (optional): lead wire

I'm giving this Whistler version separate treatment because it varies somewhat from the basic Whistler. Although it was originally created as a deepwater striped bass fly, it has become a standard tarpon design for muddy or cloudy Central American rivers. If you plan to tie flies for a visit there, this design is a good starting point. I show one version that Lefty Kreh taught me before my first trip to fish the Rio Colorado in Costa Rica. These flies are quite heavy. In addition to large bead chain eyes and large, heavy hooks, they often have lead wire wrapped around the hook shank underneath the body material. The intention is to create a bulky fly that pushes water, creating vibrations that help fish locate the fly in murky waters where vision is limited. These also make excellent striped bass flies when conditions call for a bulky, heavy fly. You can control the weight by adjusting the amount of lead wire used, the bead chain eyes, and the hook weight.

Tying procedures

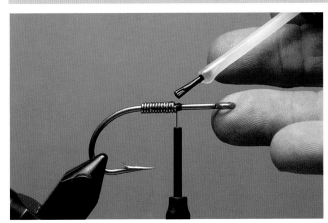

1. Wrap several turns of lead wire around the hook shank. Attach thread in front of the wire. You can coat the wire and thread lightly with head cement or, better yet, cover the lead with foam wraps, as described in the "Tying Tips" section of chapter 3.

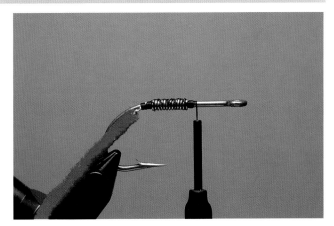

2. After cross-wrapping the wire with thread to secure it in place, strip fluff from the end of a piece of chenille, and attach the core to the shank behind the lead. Take the thread forward.

116 ■ TARPON FLIES

3. Wrap the chenille forward, and tie it down.

4. After trimming off the excess chenille, tie a bunch of bucktail on the bottom of the shank.

5. Tie another bunch of bucktail on the top of the shank, spreading the hair partially around the shank.

6. Match, moisten, and tie in four saddle hackles, curving outward on each side of the fly.

7. Tie in some Krystal Flash, Flashabou, or other tinsel on the top.

8. Add a set of large bead chain eyes on top of the head.

Additional tying notes and variations

Bright flies featuring orange, red, and yellow are a routine alternative to darker-colored Deep Water Whistlers for tarpon.

Simplified versions, using bucktail and marabou, are also quite popular.

Fishing tarpon flies

The key to successful presentation is gauging the sink rate of the fly and the speed of the fish. You must lead the fish, so that the fly sinks to the fish's level, in front or off to the side, just as the fish gets to it.

Tackling large tarpon with a fly rod is one of fly fishing's greatest challenges. This fish took a Tarpon Bunny in the Florida Keys.

Outfitter Miguel Encalada and angler Harry Robertson display a small tarpon caught on a Toad Fly in Campeche, Yucatan, Mexico. The Toad Fly is one of the most effective tarpon patterns to come along in recent years.

CHAPTER 9

Fishing the Flies Knots and Retrieves

KNOTS FOR ATTACHING FLIES TO LEADERS

Fly fishers use a wide range of knots for various purposes, such as splicing lines, attaching backing, or building leaders. Here, it's appropriate that we limit ourselves to connecting flies to leaders, since this may have an effect on the fly's behavior. Flies may perform differently, depending on the knots used. The action of the fly, or sometimes ease or quickness of tying, may dictate a choice of knot. Also, monofilament nylon and fluorocarbon sometimes don't perform similarly with the same knots. Then again, anglers simply have individual preferences. While many knots are suitable for this purpose, I've given here a select few as a starting point. They are the knots I use most, which have proven most reliable for me. Practice tying them and decide which you prefer. After all, if you can tie just one good one, you're ready to go fishing. Remember that a knot that *you* can tie well is what counts. The "best knot" is no better than the "worst knot," if it isn't tied properly. So practice these or any others you prefer, until you can tie them properly and consistently.

The Clinch Knot

The clinch knot may be the most-used knot by fishermen worldwide. Some prefer a version called the "improved clinch knot," which involves one additional step to that shown here, but I have discontinued using it, because my friend E. Richard Nightingale, engineer and author of *Atlantic Salmon Chronicles,* has convinced me that the plain clinch, when tied properly, is a superior knot. I rely on his findings, since I know of no one who has tested these knots more thoroughly. Since abandoning the improved version and resuming use of the regular clinch knot, I have never had a failure. A common error is using too few turns when making the clinch, thus allowing it to slip. When using leader material about .010-inch diameter (approx. 10 to 12 pound, depending on the manufacturer), make seven turns for an optimum connection, and don't tighten by pulling on the tag end. Lefty Kreh offers this additional important tip: Pull the tag end lightly to remove all slack so that it lies flush against the wraps around the standing line, and then pull the standing portion of the line (the part going back to your line) to tighten it. With heavier material, you'll have to use fewer turns—six or five—since it becomes increasingly difficult to tighten the knot with more turns. When using 80- or 100-pound material for a tarpon shock leader, a 3$^{1}/_{2}$-turn clinch knot is commonly used.

1. Pass the tag end of the leader through the hook eye, return it and wrap it seven times (six or five for heavier materials) around the standing portion of the leader.

2. Pass the tag end through the loop that was formed at the eye. Lightly pull on the tag end to remove all slack and lay the tag end close to the wraps.

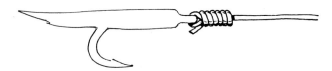

3. While holding the tag end, moisten the line and pull on the standing line to tighten the knot.

The Orvis Knot

This knot is a particular favorite of mine. In reality it is an improved or modified figure-eight knot, which has long been used for attaching braided wire to a hook. This knot is strong, quick, and simple to tie, uses little material, and is compact. One additional feature sets it apart and is appealing to anglers: When you pull the knot tight, it sets with a hitch; it snaps or clicks upon tightening. This gives great confidence, assuring that you have tied it properly. If you don't feel that little click, and see the tag end jump, retie it.

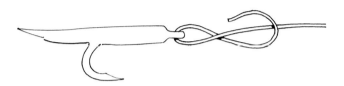

1. Pass the tag end through the hook eye and bring it back and around behind the standing line.

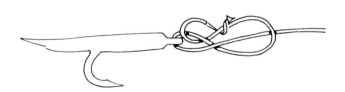

2. Bring the tag end up and pass it, from the back, through the loop that has formed.

3. Now pass it twice around the second loop you have formed, starting from the far side and pull it tight.

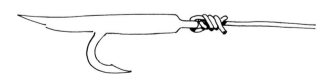

4. Now hold the hook and pull the standing end. The knot will slide tight against the hook eye. You will feel and see it seat by hitching or snapping into place.

The Non-slip Mono Loop Knot

More and more, saltwater fly fishermen choose to use a loop knot, which allows the fly more freedom of movement. Also, a knot fastened snugly to the hook eye, particularly when fishing a very light fly and employing heavy leader material, may pull to the side, off center, and affect the fly's movement. Lefty Kreh popularized this one. A word of caution: Be very attentive to tightening this knot completely. Anglers tend to be casual about seating it all the way, but when finished properly, it's one of the simplest and strongest loops you can use. It is currently the preferred loop of most fly fishers I've met. The tendency is to make the loop too large, but with practice you can consistently make it about a quarter inch.

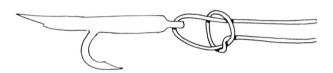

1. Tie an overhand knot in the standing portion of the leader; then pass the tag end through the hook eye and back through the overhand loop, going back *the same side it exited*.

2. Make seven wraps (or six or five as with the clinch knot) around the standing line.

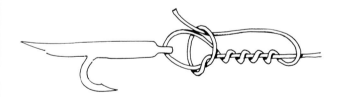

3. Pass the tag end back through the overhand loop, again going back *the same direction it exited*. Pull the tag end lightly to remove slack and draw the wraps together (as with the clinch knot).

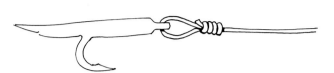

4. Moisten and steadily tighten by pulling on the standing end. The loop will remain open, not tightening against the hook eye.

The Haywire Twist

Although many fishermen prefer braided wire, with or without nylon coating, after years of testing and fishing with every sort of wire, I have limited myself to single-strand, brown stainless steel wire. I find it the most secure method. For bluefish, I use a bite tippet about four inches long of No. 9 or 10 wire, which test in excess of 100 pounds. While the strength isn't required, it doesn't readily kink as lighter No. 3 or 4 wire does. If it gets bent, you can easily straighten it with your hand and continue fishing. Light wire must be replaced when kinked or twisted, as it interferes with the fly's action. Of course, it is easier to make a haywire twist with lighter wire, so practice with that first. Some anglers find No. 6 or 7 wires an acceptable compromise. I've never determined that a short trace of the heavy wire spooks fish. Put another haywire twist on the opposite end of your bite tippet and attach your mono leader to it with a clinch or similar knot, as if connecting to the hook eye.

Fishing for barracuda and other large-mouthed, toothy fish calls for a longer wire bite leader, perhaps 12 inch. In this case, I recommend that you switch to flexible braided wire. While a short trace of single strand doesn't impede leader turnover when casting, longer sections can. Some of the newer thin and flexible wires, such as Tyger wire, can be tied with the same knots used for mono or braided lines.

1. Pass the wire through the hook eye and bend it back so it crosses the standing part in an X, not just wrapped around the standing wire.

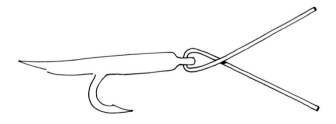

2. Hold the loop (use pliers if necessary, especially when working with heavy wire), and make three or four additional X twists.

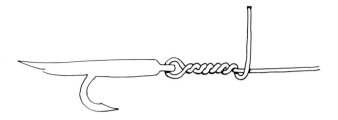

3. Bend the tag end of the wire until it is perpendicular to the standing wire and make four neat, tight barrel turns around the standing wire.

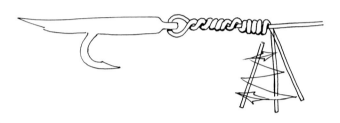

4. Bend the last few inches of the tag end to form a small handle or crank at 90 degrees to the direction of the standing wire and, while holding the haywire twist tightly, rock the tag end back and forth until it breaks off cleanly. *Never* cut it off with wire cutters. This will leave a sharp burr that will slice like a razor.

A few additional tips: It's important that you moisten your monofilament or fluorocarbon knots (saliva works well) before seating them, and that you pull slowly and steadily until they tighten. And don't skimp on the material when tying. Leave a few extra inches with which to work. Trim your knots fairly close. If a knot is tied properly, it will not slip; you don't need a long tag at the end. Finally, avoid heating the end of a knot with a lighter or match as some anglers do, believing this will prevent slippage. This will contribute nothing to the knot and will actually weaken the material. Instead, practice tying your knots properly.

RETRIEVING FLIES

Fish respond to different retrieves at different times, but this can involve more than just a change of speed. Some flies simply react differently and work better with a different motion. Therefore, it's bad strategy to rely on one motion and tempo when fishing your flies. While many anglers use a one-hand line strip exclusively, for the vast majority of my saltwater angling along the east coast, I employ a hand-over-hand retrieve. This is not to say that using a two-handed retrieve is superior to a single-handed retrieve. Nor is the reverse true.

Every angler should be proficient with both techniques and use them as the fish or the fly dictates. A Clouser Minnow generally works best with long one-hand strips, followed by a pause to allow the fly to sink. Bonefish flies usually produce best with shorter, staccato strips. The strip-strip motion is also effective for tarpon. However, Siliclone flies call for a steady motion, which the intermittent stop-and-go action of one-hand stripping can't produce. You should regularly experiment with retrieve speeds and alternate between one- and two-hand retrieves.

One-hand retrieve sequence

Drawing the line repeatedly over the first or second finger of the rod hand can be accomplished slowly or quickly, making long or short pulls, but it always gives the fly an intermittent, stop-and-go action.

Two-hand retrieve sequence

After casting, tuck the rod under your arm (right-handers generally put the rod under the right arm) and take the line in either hand. The two-hand technique is more versatile, in that it can duplicate the stop-and-go motion of the one-hand strip but gives other options too. Alternately pulling the line with each hand not only makes a high-speed retrieve possible, but it also can be used to give the fly a very slow, steady swimming motion. Hooking a fish is also easier because one hand is always pulling the line and the rod tip is pointing toward the fly. A sharp tug on the line is generally all that is needed to set the hook. Strikes are often missed during the pause of a one-hand strip.

RECOMMENDED READING

Bay, Kenneth. *Salt Water Flies*. Philadelphia: Lippincott, 1972.

Clouser, Bob. *Clouser's Flies: Tying & Fishing the Fly Patterns of Bob Clouser.* Mechanicsburg, PA: Stackpole Books, 2006. Includes further information and details about all the Clouser flies.

Fernández, Chico, and Aaron J. Adams, Ph.D. *Fly-Fishing for Bonefish*. Mechanicsburg, PA: Stackpole Books, 2004. Provides full treatment of fishing the flats for bonefish.

Jaworowski, Ed, and Bob Popovics. *Pop Fleyes.* Mechanicsburg, PA: Stackpole Books, 2001. Detailed treatment of epoxy flies and drying wheels, further discussion of all the other Popovics flies contained in this book, extensive treatment of the subject of silicone and its use for various Siliclone flies, details about the Surf Candy, and more.

Sosin, Mark, and Lefty Kreh. *Practical Fishing Knots.* Guilford, CT: The Lyons Press, 1991. For more information and additional knot options, I recommend this book in particular.

INDEX OF FLY PATTERNS

Those flies indicated in **bold** are illustrated in step-by-step sequences in the text.
Those listed in standard type are shown only as finished flies.

Alba-Clouser	38	Ka-Boom-Boom	87
Apte Tarpon Fly	107	Keel Eel	32
Argentine Blonde	33	Lee Cuddy	93
Baby Tarpon Flies	109	Keel Fly	32
Bendback	30	Long-legged Crab	68
Black Death	108	**Mini Puff**	99
Bob's Banger	77	Palolo Worm	75
Bonefish Clouser	38	Platinum Blonde	34
Bonefish Slider	105	Pop Lips	53
Bonefish Special	92	**Popper**	86
Bucktail Deceiver	35	Potomac Popper	88
Bunny Fly	110	**Rabbit Eel**	72
Chernobyl Crab	64	Ron Dong Crease Fly	80
Chinese Claw	108	Schoolie	58
Cinder Worm	74	**Seaducer**	49
Clouser Brightsides Minnow	83	**Siliclone**	51
Clouser Deep Minnow	37	**Slab Side**	54
Cockroach	112	**Slider**	89
Crazy Charlie	94	Slim Jim	90
Crease Fly	79	**Snapping Shrimp**	101
Deceiver	40	Squid Deceiver	41
Deep Candy	58	Spouter Popper	88
Del Brown's Permit Fly	66	Strawberry Blonde	34
Dink Popper	88	Super Swimming Shrimp	104
Floating Minnow	82	**Surf Candy**	56
Frankee Belle	93	**Swimming Shrimp**	103
Glass Minnow	31	Tarpon Glo	108
Gotcha	96	Tarpon Toad	114
Gurgler	84	**Toad Fly**	113
Half and Half	43	**Ultra Shrimp**	69
Hi-Tie	45	**Whistler**	59
Homer Rhode Streamer	108	**Whistler for Deep Water/Tarpon**	115
Honey Blonde	34	Wide Side	32
Horror	98	Wiggle Worm	75
Integration Blonde	34	**Wobbler**	61
Jiggy	47		